Joyeux anniversaire, Mon Mam,

J'espère que tu appréciez ce livre!
Je crois que tu le trouverez
intéressant. À l'intérieur du livre
est un papier avec un autre cadeau~

Nous partagerons beaucoup plus
d'anniversaires spéciaux ensemble!

Je t'aime beaucoup toujours!!

Ton Petit Papillon

CREATIVE EVOLUTION

Also by Amit Goswami

The Self-Aware Universe

Quantum Creativity

The Visionary Window

Physics of the Soul

The Physicist's View of Nature

The Quantum Doctor

God Is Not Dead

CREATIVE EVOLUTION

A Physicist's Resolution between Darwinism and Intelligent Design

AMIT GOSWAMI, PH.D.

QUEST
BOOKS

Theosophical Publishing House
Wheaton, Illinois • Chennai, India

Quest Books
Theosophical Publishing House
P.O. Box 270
Wheaton, IL 60189-0270

www.questbooks.net

Cover image: © Craig Stevens/Images.com

Library of Congress Cataloging-in-Publication Data

Goswami, Amit.
Creative evolution: a physicist's resolution between Darwinism and intelligent design / Amit Goswami.—1st Quest ed.
    p.   cm
Includes bibliographical references and index.
ISBN 978-0-8356-0858-9
1. Evolution (Biology)—Philosophy. 2. Intelligent design (Teleology) 3. Quantum theory. 4. Religion and science.
I. Title.

QH360.5.G67 2008
576.8—dc22   2007052718

5   4   3   2   1   *   08   09   10   11   12

Printed in the United States of America

CONTENTS

*I*LLUSTRATIONS

PREFACE

This book is about a new integrative biology that has the potential to integrate all of biology's many controversies, solve its paradoxes, and explain the many anomalous data that befuddle biologists, especially the Darwinists. In my earlier books, I have used the ideas of quantum physics and the primacy of consciousness to integrate the various forces of psychology, such as psychoanalysis, behaviorism, and transpersonal psychology; to integrate science and spirituality; and to integrate conventional and alternative medicine. This book is the final piece of integration. Some of the material developed earlier is repeated but always framed in the new context. In the main, though, this book presents many new ideas that, I believe, provide the rudiments of a new paradigm in biology. With the publication of this book, I can truly say that a framework exists for integrating all science into a coherent whole, an endeavor that can be called *science within consciousness*. Most importantly, this book provides a resolution of the controversy between evolutionism and intelligent design.

Echoing the psychologist Carl Jung, we may ask whether science, like all other human endeavors, can be a search for the soul. Yes, it can. However, biology as practiced now could not be farther from such a search. That is why so many people find it doesn't satisfy their souls.

So who is the audience for this book? By avoiding the jargon of both biology and quantum physics as far as practicable and by keeping the explanations simple, I have presented creative evolution as an entirely user-friendly new biology for nonspecialists. Yet make no mistake about it: I present a paradigm-shifting idea for biology, and professionals are invited to engage this idea as well.

If you are an established professional who is tired of seeing the field of biology become more and more preoccupied with engineering questions while shunning the search for its own soul, let alone the soul of humanity, then I think that this book will contribute toward enriching your worldview.

If you are an aspiring biologist who is tired of the biology hierarchy telling you what to think, and one who is quite aware of the shortcomings of molecular biology and especially of Darwinism, I hope this book encourages you to continue thinking within this new paradigm.

If you are a believer in intelligent design but see through the naïveté of creationists and are genuinely curious whether intelligent design can ever be truly incorporated in a full-blown scientific paradigm, this book will satisfy you.

If you don't care for intelligent design talk but see through the naïve materialism of neo-Darwinist thinkers, I hope this book will provoke you and make you think further.

And if you are a true believer (either in the Bible or in scientific materialism), I have this to say to you. In the sixties, I was impressed with a TV commercial for a certain cereal. The cereal was intended for strengthening little kids so that they would no longer be afraid of bullies. But what if the bullies also strengthen themselves by eating the same cereal? The ad assured its viewers that the cereal would not be sold to bullies. It was an exclusive product, meant for the benefit of a select few! But this book is not about a cereal; it is about science. I believe, like the psychologist Abraham Maslow, that science has no entrance requirement: no exclusivity permitted. So you, too, as a believer, are warmly welcomed to the debate.

With this book, a vision that took root in the world in the 1970s with the publication of Fritjof Capra's *The Tao of Physics*—the vision that science and spirituality can be integrated—has now come to fruition. My sincere hope is that this integration will act as a balm to heal not only the deep division signified by the struggle between the evolutionists on the one hand and the creationists and proponents of intelligent design on the other, but also all deep divisions in our society and culture arising from the division of science and spirituality. And above all, I hope that the seed of a new biology proposed here will inspire many generations of students to develop a mature new field: biology within consciousness.

In the early days of this research I received much inspiration from several biologists: Dick Strohman, Mae Wan Ho, Beverly Rubick, and especially Rupert Sheldrake. I thank them all. I especially thank biologist Dennis Todd for introducing me to the subject of directed mutation. I also thank my wife Uma Krishnamurthy for her continuing focus on the importance of emotions in our lives and evolution. Finally, I thank my editors Sharron Dorr and Jane Andrew, and the staff of Quest Books for bringing out a fine publication.

PART I

*I*NTRODUCTION

1

God *and* a New Biology

The Overview

*D*arwin's theory of evolution is the foundation of biology, but every modern biologist—in moments of total honesty—hears the foundation creaking. Darwinism is a theory of continuous evolution. But it's now an open secret that fossil gaps—discontinuities in evolutionary fossil lineages—pose a serious threat to the complete validity of Darwin's theory. It is also well known that Darwin's theory and God's existence are mutually exclusive ideas. But if Darwin's theory is at best an incomplete theory of evolution, only able to explain its continuous epochs, there's room for God to make a comeback.

Intelligent design theories try to revive God, either explicitly, as in creationism, or implicitly, by pointing to the intelligence and leaving us to infer a designer, but end by denying evolution altogether. What an ingenious way to sidestep the fossil gaps: no evolution, no fossil gaps to explain. Unfortunately, too much credible evidence exists in favor of our evolutionary ancestry for this dodge to work.

But should we throw the baby out with the bath water? Is there any substance in intelligent design theory (let alone creationism) that warrants serious scientific attention? Do these theories present any ideas with which biologists must come to terms? The unprejudiced scientific answer to both questions is yes. And the important idea I am talking about is this: An intelligent design of life suggests that biology must come to terms with the feeling, meaning, and purposiveness of life and with the idea of a designer.

The current materialist basis of biological theories, including Darwinism (and its more recent revision, neo-Darwinism), prevents these theories from properly including these theologically tinged ideas. In this book I demonstrate that including the idea of creativity in biological evolution reconciles the notions of evolution with those of intelligent design by a purposive designer. In fact, I show that evolution proves intelligent design. Furthermore, when the question of the purposiveness of life's design and the existence of the designer are reconciled with the evolutionary ideas of Darwinism by using quantum physics and the primacy of consciousness, many other paradigmatic difficulties of biology and biological evolution are also resolved.

Although intelligent design theorists miss it, one piece of compelling experimental evidence exists for design and purposiveness of life: The evolution of life proceeds from simplicity to complexity. By looking at the fossil data alone, any intelligent person can distinguish between time past and time future. In other words, the fossil record of biological evolution gives us an unmistakable arrow of time. The intelligent design theorists miss the importance of this fact because of their contention that there is no evolution at all.

Darwinists, on the other hand, do attempt to understand the evolutionary trend to develop complexity and intelligence. But their attempts are based on the idea of genetic determinism—that evolution is determined and driven by genes' need to survive (Dawkins 1976). This idea enables biologists to attribute all sure signs of the intelligence of life—feeling, meaning, and indeed consciousness itself, just to name a few—to adaptive epiphenomena of the genetic drive to survive environmental changes. The notion is very weak on two scores. First, compelling theoretical arguments have been presented showing that the molecules of which genes are part do not have the capacity to process feeling, meaning, or consciousness. How then can such qualities evolve adaptively from nothing? Second, most biologists believe that biology at its most basic is connected to physics, but in recent years physics itself, under the pressure of compelling experimental data, has abandoned strict determinism and made room for occasional conscious choice. Despite this ideological change, physics has escaped major revision because it concerns itself with behavior en masse: It rests on a statistical determinism that holds for a large number of objects or a large number of events, or both. The biologist has no such consolation, because in biology the behavior of a single organism is as much a concern as that of the many.

In short, biology must reconcile itself with the revolutionary aspect of quantum physics: that is, indeterminacy and choice by a quantum consciousness. I show in this book that such a reformulation of biology puts the idea of creative evolution on a firm footing. With such a footing, a theory of creative evolution can not only integrate such disparate ideas as intelligent design and evolutionism or discontinuity and continuity, but also unify ideas of development (how biological form arises from a one-celled zygote) with ideas of evolution.

The sympathy that a portion of the American public feels for intelligent design theory is not merely of religious origin. It can also be traced to an uneasiness about the attitudes implicit in Darwinian evolutionism, and indeed in scientific materialism itself. How can we take these viewpoints seriously when they denigrate our intelligence, our capacity to process feeling and meaning, and our consciousness itself by naming them a meaningless, epiphenomenal dance of elementary particles and their conglomerates, the genes? We are also uneasy because Darwinism tells us nothing significant about the future of our evolution. Does evolution lead to increased intelligence? Darwinism is equivocal: Evolution can lead to more complex or less complex organisms, more intelligence or less. We cannot predict; the outcome is left to chance and survival necessity.

The theory developed in this book, creative evolution, is unequivocal. Creative evolution is geared toward higher and higher intelligence, toward developing qualities of intelligence that our religions and spiritual traditions identify as godly. If you hear in that statement the echo of the ideas of such philosophers as Sri Aurobindo and Pierre Teilhard de Chardin, that is no coincidence. Creative evolution incorporates the revolutionary ideas of these two great thinkers.

In addition to making room for such ideas, the inclusiveness of creative evolution permits resolution of several long-standing sore points in biology. For example, the ideas of Lamarckism—that traits acquired in an individual lifetime can be inherited by offspring—are reintegrated into biology, ending a long controversy. This reconciliation is accomplished in the context of a much-needed explanation of instincts. The new biology proposed here incorporates a proper resolution of the mind-brain problem: Specifically, it presents a paradox-free treatment of the neurophysiology of perception. Most importantly, an integrative reformulation that locates biology within consciousness enables us to arrive at a proper formulation of biology that includes feeling and that can begin to treat heterogeneity, the individual differences between organisms. Furthermore, locating the new biology within consciousness leads to a satisfying bioethics and a deep ecology. It also gives us a new perspective for dealing with the issue of survival after death, a perspective that opens the door for additional reconciliation of biology and religion.

Apropos of religion, a word about the title of this chapter, which links God and biology. I freely use the terms *God* and *quantum consciousness*—the causal source of conscious choice in quantum physics—as interchangeable concepts, an equivalence I explore further in chapter 2. This choice to use the term *God*, made as a gesture toward the viewpoint of the faithful believer, should not be taken as an affront to the scientific sensitivity of the professional biologist. As you will see, the God of this book is an objective organizing principle. Seen as a new, objective organizing principle, the idea of God becomes useful under the most stringent qualifications as an element of new science.

So Where's the Problem?

Many people dismiss the idea of intelligent design offhand because "everybody knows" that Darwin and his followers have shown evolution rules out intelligent design and a designer. It is true that Darwin's theory attempted to explain evolution without invoking the concept of intelligent design. However, it is also true that, according to Darwin's theory, evolution is continuous and should produce a continuous fossil record of all evolution. Unfortunately, the fossil records show glaring gaps at many important junctures. In other words, evolution is not only continuous but also discontinuous (Eldredge and Gould 1972). Evolution has been compared to punctuated prose: The punctuation marks are discontinuities in otherwise continuous text. Darwinism cannot provide a fully credible explanation of such discontinuity. In this book, I take the discontinuity in biological evolution seriously and show that, like the well-known discontinuous jumps of our own creative experiences (Harman and Reingold 1984), the fossil gaps are signatures of biological creativity. And creativity is a definitive sign of intelligence. In this way, I show that evolution proves intelligent design.

However, creativity and intelligent design also imply a creative and intelligent designer, or, as I term it, a nonphysical and nonmaterial organizing principle. Such organizing principles have been proposed in biology from time to time, but until now it has not been clear how this kind of organizing principle could operate within a scientific framework. In this book I will show that some recent developments in quantum physics and elsewhere are telling us how much-needed nonphysical and nonmaterial organizing principles can be incorporated in biology.

Every biologist must be painfully aware that biology is an incomplete science. It needs new organizing principles, ones that are nonphysical and nonmaterial, to explain three perennial mysteries: the difference between life and nonlife (Davies 1988), the development of an embryo into an adult biological form (Sheldrake 1981), and, as emphasized here and by Eldredge and Gould (1972), the discontinuous epochs of evolution. Unfortunately, it is not politically correct for a biologist to admit these shortcomings in public. In this book, I will show that the introduction of new nonphysical and nonmaterial organizing principles (yes, principles, plural; we need more than one) can complete biology as a science. In this way I will set a framework from which biologists can work to rid their field of the paradoxes, controversies, and anomalies that have plagued it from its very inception to the present, including the highly politicized controversy pitting evolutionism against creationism and theories of intelligent design.

THE ORGANIZING PRINCIPLES OF OLD BIOLOGY

The current biological paradigm (the "old" paradigm) is based on two organizing principles. One is the principle of *upward causation*: All biological phenomena arise from the interaction of microscopic constituents of matter called *molecules* (and ultimately from the interactions of submicroscopic particles called *elementary particles*). This principle assumes a molecular basis of life, the idea that life can be reduced to the movement of molecules. This assumption has given us molecular biology—a science proclaiming that all the functions of a living cell and conglomerates of cells can be understood in terms of the physics and chemistry of molecules, especially large "macromolecules" called *proteins, DNA* (deoxyribonucleic acid molecules), and *genes* (portions of DNA). In biology, the dogma of upward causation is expressed as genetic determinism: Genes determine all biological form and function.

The other organizing principle of old biology is that *evolution is determined by chance and necessity*. This principle, discovered by Charles Darwin in 1857 (Darwin 1859), forms the basis of the evolutionary model called *Darwinism* (see Mayr 1982 for a history) that is accepted, explicitly or implicitly, by most biologists. According to the latest version of Darwinism, evolution proceeds in two stages. The first stage is the chance production of variations in the hereditary components of life (the above-mentioned genes). The second stage is selection from among these variations, dictated by the necessities of survival for species coping with changes in the natural environment. This process is called *natural selection*. The genetic changes that cause the beneficial chance variations occur rarely, but working over millions of years this slow, two-step Darwinian mechanism accounts for all facets of evolution, according to most biologists.

However, both of these organizing principles are mired in controversy. A debate continues about whether molecular biology can ever explain what life is or how it originated. After some initial successes, the molecular synthesis of life in the laboratory has remained elusive (Davies 1999). Controversy also swirls around theories of development, that is, how a single-cell zygote develops into a full-fledged form, the organism. Is organismic development solely the handiwork of upward causation from the genes, or is there a role for the environment (Goodwin 1994)? Might there even be a role for new organizing principles in explaining the full intricacies of development (Sheldrake 1981)?

Perhaps it was such controversies that prompted this remark by the biologist Brian Goodwin (1994): "I don't think biology at the moment is a science at all, at least in the sense that physics and chemistry are sciences. We need to know the universal ordering principles just as Newton provided them for the inanimate world."

The most publicly visible controversy is, of course, about evolution. The main scientific evidence for biological evolution is the fossil data. According to Darwinism, the story of evolution is a continuous one: The transition from an earlier species to a later one is incremental and continuous, and the fossil data should reflect that. Unfortunately, this premise is not borne out; we find the famous fossil gaps already mentioned, gaps that appear when the fossil data are viewed as a chronology of evolutionary ancestry. Darwin himself knew about this problem, but he was optimistic, justifiably, that further investigation would turn up intermediates to fill the gaps. Indeed, we do occasionally hear about discoveries of intermediates, but according to Darwin's theory, thousands upon thousands of these intermediates should have been discovered by now. Such discoveries have not happened. So the fossil gaps raise legitimate doubts about the veracity of Darwinism (and its later incarnation, neo-Darwinism) as a complete theory of evolution. In science, we must take data seriously, and by now much research should have been carried out toward replacing Darwinism. But this has not happened either; the challenge to Darwinism (and neo-Darwinism) within biology has been sporadic, and even these sporadic efforts have, to a large extent, been ignored (see chapter 10 for a bit of history). Because biologists in the main have been less than candid about this matter, the challenge to Darwinism was taken up outside the scientific arena and has become highly politicized.

What Is Intelligent Design?

In the public arena, the challengers who have gotten the most visible support are those who challenge the idea of evolution itself. Suppose there is no evolution; it is a fact that the fossil data indicate much stasis; many organisms seem not to change for long periods of geological time. The challengers posit the following: Suppose that, instead of being the result of evolution, all life is the result of intelligent design by a designer who acts all at once. Certain biological forms are too complex to have originated through chance and necessity, these challengers maintain.

Some intelligent design theorists resort to an old philosophy called *creationism*, following the Genesis chapter of the Old Testament of the Bible (Gish 1978). This theory flatly declares that God created the world and all the biological species six thousand years ago in six days: There is no evolution.

The idea of creation by God is an aspect of God's *downward causation*. The term reflects the tendency to picture God as an emperor sitting on a throne "up there" in the heavens, brandishing the causal wand of downward creation in His hand. Such anthropomorphic pictures of God irritate scientists (and probably many nonscientists as well).

The resurgence of creationism as an alternative to evolutionism has increased the stakes, because the context of the controversy has reverted to the old struggle for "worldview control" between science and the Christian church, and so a lot of negative emotion has been generated. Scientists feel invaded by theology: How can it be called science when an idea of faith (the biblical God) is brought to bear on science? On the face of it, creationism does sound unscientific, even to an unprejudiced reader, because of its biblical origin; it is true that the validity of the Bible is based on faith, not experimental data.

Can religion be taken out of this debate? More recently, some serious scientists, among them some professional biologists, have begun positing the idea that species are created by an intelligent designer—maybe God, but it is kept implicit—without subscribing to the biblical baggage. In scientific language, this "causal creation by the intelligent designer" is just another organizing principle, albeit a nonphysical and nonmaterial one.

Any perceptive person can see design in life. Can that intelligent design come from linear, step-by-incremental-step chance and necessity? In spite of all the time available for chance and necessity to do their thing, detailed reasoning and probability calculations by the intelligent design theorists and their sympathizers (Shapiro 1986; Behe 1996) raise legitimate doubts about the validity of Darwinism as a mechanism for producing the complicated, often nonlinear designs that life exhibits. These doubts certainly justify scientific consideration of alternatives to Darwinism. If such an alternative involves the introduction of additional organizing principles in biology, so be it.

A CLASH OF WORLDVIEWS

In 2005, the Kansas state board of education got a lot of attention in the media because they adopted a school curriculum that favors the teaching of intelligent design along with evolutionism. Two national science organizations, the National Academy of Sciences and the National Science Teachers Association, started black-balling the state of Kansas by refusing to allow it the use of key science education material. In official statements, the organizations said they were disappointed that the Kansas board deleted a paragraph in an earlier draft of the education standards in which science was defined as "a search for natural explanation of observable phenomena." Sounds fair, doesn't it? But in an interview Jay Labov, a senior advisor for education at the National Academy of Sciences, gave away the real concern, which was that "the deletion could lead students to believe that supernatural explanations also may fall within the purview of science" (Weiss 2005).

These scientific organizations are implicitly defining nature as consisting of the material universe. Any organizing principle that is nonmaterial is automatically excluded from science by definition.

However, mainstream scientists themselves, biologists included, have a fundamental but unproven metaphysical assumption behind their work called *scientific materialism*. This philosophy posits that when everything is said and done, all things and phenomena of the world can be understood on the basis of only one causal substratum—the elementary particles, the basic building blocks that make up matter. Can one unproven metaphysical assumption be allowed to exclude other metaphysical assumptions in science? Isn't that akin to the very faith that scientists complain about when they criticize creationism and intelligent design theories?

You will often hear mainstream scientists declare that, in this technological age, the validity of scientific materialism should be obvious and that it is "absurd" to speak of nonmaterial organizing principles. That way lies dualism, the idea that two separate and irreducible principles (in this case, material and nonmaterial reality) can coexist. Such scientists point to the perennial logical challenge to dualism: Two entirely different substances with nothing in common cannot interact, cannot impose causation upon one another. How does a nonmaterial designer interact with matter to design something? Is there a mediator of their interaction? If not, how can they possibly interact?

For contrast, let's look at the causal picture of upward causation that is supported by most scientists. Elementary particles make atoms, atoms make molecules, molecules make living cells with those all-powerful genes, some of the living cells (guided by the genes) make the brain, and the brain makes all subjective experiences, such as consciousness, thoughts, feelings, and so on. Cause rises upward from the elementary particles, and all causation is upward causation.

Dualism is scientifically absurd, but is this picture of the design—humankind as the product of upward causation—any less absurd? Can you really believe that all your thoughts and meanings, your feelings and struggles with values, and indeed your consciousness itself, are the results of a random dance of elementary particles or genetic determinism? That you are a purely ornamental epiphenomenon, a secondary consequence of the random movement of the matter in your brain cells? Even the scientific proponents of the idea of purely upward causation don't really believe that! If they themselves are merely causally impotent epiphenomena, why do they take themselves and their ideas so seriously?

We see that the great evolution debate between science and Christianity is really a clash between two worldviews, both of which are faulty. Is there a way out of this dilemma?

WHAT DO THE DATA SAY?

In science, experimental results are the final arbiter; if data falsify the predictions of a theory, we must give up the theory, or at least suitably modify its scope. So let us look at data.

Biologists claim that creationism does not stack up well against that ultimate scientific test. This claim is correct. In creationist theory, God created the world six thousand years ago in just six days. This statement has been falsified beyond doubt; much geological and even physical data (radioactive dating) exist to show convincingly that the Earth is about five billion years old.

But the creationists make an equally valid claim that Darwin's theory of evolution is falsified because of the fossil gaps. One of Darwin's major theoretical predictions was that gaps would eventually fill up as we perfect our empirical investigations; many later biologists have expressed similar optimism. Well, we have perfected the techniques of empirical investigation, and just as the age of the Earth can today be stated accurately, so can one state accurately that the fossil gaps are mostly real: They're here to stay.

To be sure, a few intermediates have caused a stir. For example, in reports of intermediates by the biologist J. G. M. Thewissen and his collaborators (1994), much is made of an intermediate fossil found for an animal that could move both in land and in water, a land-walking whale, so to speak. But how many such cases exist today? A thorough search of the Internet yields only about fifty cases of intermediates in the entire fish-amphibian-reptilian lineage of about forty-two thousand species.

The discovery of intermediates is important because it discredits creationism in favor of evolutionism; unfortunately, evolutionism is not the same thing as Darwinism. I repeat: According to theoretical predictions of Darwinism and its later versions, there should have been thousands upon thousands of reported cases of intermediates filling up most of the fossil gaps. That hasn't happened, and therefore the question of the fossil gaps cannot be refuted simply because a few cases of transitional fossils have been found.

Because both creationism and Darwinism are based on faulty philosophy and both are falsified in part by the data, should we give them up entirely? No, there is a middle ground.

In spite of the fossil gaps, evolutionism does have a solid empirical fact on its side: Some species have so much in common that the idea of a common origin, a tree of life, so to speak, seems unavoidably obvious. Darwin's original idea about such similarities (called *homology*) has now been corroborated with very convincing data (Carroll 2005). Such a tree has gaps in it, to be sure (fig. 1), reflecting the fossil gaps. But the idea that species evolve from ancestors is too consistent with the data to give up in favor of the alternative, as presented by creationism and intelligent design theory, that God created all species all at once, independently of one another.

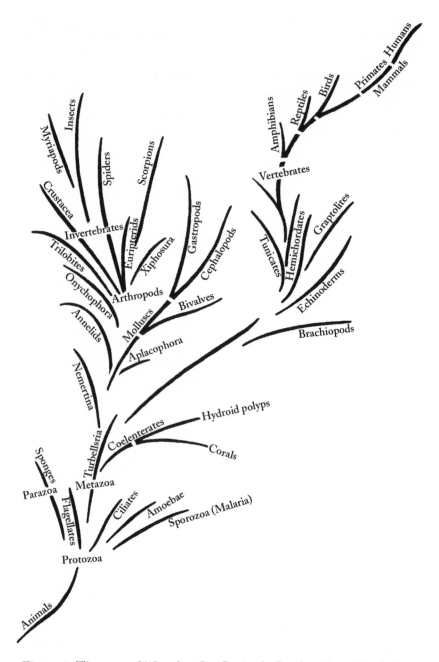

Figure 1. The tree of life, after Sir Gavin de Beer's *Atlas of Evolution* (1964). The branches and sub-branches are punctuated by (fossil) gaps.

Further support for the idea of evolutionary ancestry comes from embryology. In some species, embryos at early stages of development so clearly resemble other species that we are forced to conclude that the former must have evolved from the latter (Wolpert 1993). So creationists and intelligent design theorists miss the boat when they deny entirely the idea of evolution. "Evolution, no; God, yes" is not an empirically valid approach.

Intelligent design theorists implicitly fall prey to the Christian notion that God's design, like God, must be perfect from the get-go. Empirically, it is easy to see that God's design has many imperfections (for example, the intermediates mentioned above). Instead of making additional dualistic assumptions about reality to deal with these imperfections (such as assuming a split between good and evil), why not assume that the design is not perfect initially but rather evolves toward perfection? This choice makes evolution and God quite inclusive of each other.

The revised motto is "Darwinism, no; evolution, yes; God, yes." Even Darwinism must be given its due. It stands up as an acceptable theory for the continuous epochs of evolution; it's just not a *complete* theory of evolution.

INTELLIGENT DESIGN
WITHOUT A DESIGNER?

Darwinists, and, inexplicably, most biologists with them, resort to sleight of hand here. They do not deny the intelligence that exists in living organisms: That is too obvious to deny. Instead, they claim that the intelligence of life forms has no causal efficacy (Dawkins 1976). The intelligence of an organism is simply one way the organism's genes ensure their survival. This idea is an offshoot of genetic determinism. How then does the intelligence of an organism, its ability to process meaning, feeling, consciousness, and all that, develop? According to the Darwinist, it develops not because an intelligent designer made life with those causally efficacious qualities, but because the organism obtains a selective advantage by having them. They are adaptive emergent qualities.

This argument does not make good biology, and for a very good reason. Animal and human behavior clearly show that real intelligence with causal efficacy exists in living organisms. Furthermore, the ability of organisms, especially humans, to process meaning with creativity and purpose—as supported by much of our own experience and by objective data—indicates causal efficacy. Witness all the creative work by the many scientists of the twentieth century alone that has causally "disturbed the universe." Think of the discoveries in relativity and nuclear physics that made possible the atomic bomb—and the consequent destruction. That work cannot be dismissed as an example of genetic determinism. If the design shows both intelligence and causal efficacy, the concept of an intelligent and purposive designer cannot be dismissed offhand.

Furthermore, depending on the idea of adaptive advantage for the emergence of all things unexplainable by biology—among them feeling, meaning and purpose, and indeed consciousness itself—is no less foolhardy than the old religionists' attempt to account for everything inexplicable as acts of God. Matter just isn't up to the task of handling those things: Scientists and philosophers have been able to argue quite compellingly that matter cannot process meaning (Penrose 1989; Searle 1994) or produce consciousness (Chalmers 1995). In the most rudimentary form, their arguments are quite easy to follow:

- To see why matter cannot process meaning, consider trying to program a computer to process meaning. Computers are symbol-processing machines, so you start by reserving some symbols for processing meaning. But a little thought tells you that now you need more symbols to keep track of the meaning of the meaning symbols. This dance goes on ad infinitum. In other words, there isn't a computer large enough to process meaning.

- Matter consists of objects. Properties of complex objects can be reduced to the movement of other simpler objects. Conglomerates of simple objects can be used to explain the workings of a bigger complex object. So if consciousness were an object, certainly an entirely objective explanation could be found. But how do you experience consciousness? Does it not have a subjective component as well as an objective one? In other words, don't you experience your consciousness as a subject looking at object(s)? But an objective explanation can never be given for the subjective aspect of consciousness. In other words, materialism can never fully explain consciousness.

If matter cannot even process meaning or organize consciousness, how then can matter produce meaning and consciousness as adaptive epiphenomena from which nature may select?

SEEKING A MIDDLE GROUND

That the fossil data have gaps clearly suggests to some biologists that there are two tempos of evolution, one slow and one fast (Simpson 1944; Eldredge and Gould 1972, 1977; Grant 1977). The idea is that during the fast epochs of evolution, there just isn't enough time for fossils to form, hence the fossil gaps. In Niles Eldredge and Stephen Jay Gould's very evocative description, evolution is like continuous prose modulated by punctuation marks—commas and periods. Darwinism is a theory of slow-tempo evolution; it can only explain the continuous prose. So what is the mechanism behind the fast tempo of evolution, the punctuation marks? A mystery is created. Instead of dealing with the mystery, establishment biologists busily look for theories to explain away the fossil gaps without introducing the inconvenience of a fast tempo. No fast tempo, no new mechanism. Slow and reliable Darwinist chance and necessity will do!

Among such theories, the most successful in terms of popularity, and the one cited in textbooks, is the geographical isolation theory of the theoretical biologist Ernst Mayr (1942). Suppose that a small population of a certain species gets geographically isolated (by a mountain range, for example) for a period of time. During that period, because the environmental challenges are different, natural selection would lead to the evolution of different traits for the two populations so separated. This isolation can easily lead to a situation in which the two populations, when they come together again, can no longer mate with one another. In other words, the two populations have become reproductively isolated, which is the current definition of *speciation*, the change of one species into another. Additionally, in a small population, the gene mutation rate increases, so that a population may become reproductively isolated rather quickly.

The theory fails in several ways. Invented as a response to a specific problem, the theory is unverifiable (and thus not truly a scientific theory), simply because it is impossible to create such a scenario for empirical study. A more important question, however, is this: Can geographical isolation explain the data on fossil gaps in *all* cases of speciation? No, it cannot. It certainly is a viable explanation of speciation for what we call *microevolution*, the evolution of simple organisms involving only a few genetic changes. However, it cannot explain macroevolution, or evolution of new species involving a large number of genetic changes. The reason is subtle, as I will show next.

Probability calculations alone preclude Darwinism's ability to explain all evolution, whether micro or macro. I have cited biologist Robert Shapiro's (1986) work earlier in this chapter. Shapiro showed that the maximum number of chance events available over a billion years of evolution is 2.5 x 10^{51}. The astrophysicist Arne Wyller (2003), on the basis of very conservative assumptions, deduced that to create the billion multicellular species that have ever existed on Earth until now (according to the Harvard biologist Richard Lewontin) requires more than $10^{1000000000000}$ chance events. This figure is obviously far, far, far greater than the maximum number of chance events available as calculated by Shapiro.

Either way, we are forced to the conclusion that there must be alternative routes that complement Darwinian evolution. Mere chance and necessity cannot do it all. Biologists have to face up to this compromise.

But religionists must also recognize the need to compromise. Suppose you are willing to see the intelligent designer not as a God or emperor sitting on a throne "up there," but as a new, objective organizing principle capable of causation. As mentioned earlier, this mode of action can be called *downward causation*, but only to contrast it with the upward causation of materialist vintage. Suppose further that you realize the six days of biblical Genesis could be a metaphor for all the fast-evolution epochs that the punctuation theorists Niles Eldredge and Stephen Jay Gould have proposed to explain the fossil gaps. Finally, you acknowledge that creativity is empirically known to be sudden, instantaneous. You might now find yourself at a startling conclusion: Downward causation by God is the organizing principle for species creation during the epochs of fast tempo. During those epochs God intervenes, so to speak, and creates new species not from scratch, but out of the existing manifestations. In this view—what I call *creative evolution*—creationist and intelligent design theories and Darwinism are both vindicated, in the main. They can live together, and more. Together they can explain both tempos of evolution, fast and slow, and the main ideas of both theories are integrated.

How do the odds work out for creative evolution? This is where God's purposiveness enters in a major way. Just as a creative artist can create new work against all odds because he has a template in mind, God is able to evolve the new because She, too, has a purposive blueprint in mind (for more on this, see chapter 4).

By entertaining the idea of an intelligent designer capable of downward causation (call this designer God or not), we still face a genuine scientific problem. It can be posed this way: How does a supposedly nonmaterial designer (it has to be nonmaterial, not bound by material laws, which allow only upward causation) interact with the world of matter to design anything? This is the previously stated problem of dualism.

If the idea of downward causation were an isolated idea invented to solve the special problems of fast-tempo evolution and purposiveness of life, if it were needed nowhere else in science, then it could not be called a scientific idea, end of story. But the intriguing situation is this: The idea of a God as an agent of downward causation has emerged in quantum physics (Goswami 1989, 1991, 1993, 2000, 2002; Stapp 1993; Blood 1993, 2001) as the only legitimate explanation of the famous observer effect. (Readers skeptical about this statement should see these original references, especially Goswami 2002.)

In quantum physics, objects are depicted as possibilities (a possibility wave); yet when an observer observes, the possibilities collapse to an actuality (the wave collapses to a particle, for example). This is the observer effect. Most importantly, quantum collapse of possibility into actuality is discontinuous, so the discontinuity of the punctuation marks of evolution is instantly accommodated if we see them as instances of quantum creativity—discontinuous collapse of quantum possibility into actuality (Goswami 1997a). This premise forms the core of the theory of creative evolution.

Most importantly, downward causation introduced via quantum collapse is consistent with a philosophy, *monistic idealism*, that avoids dualism and transcends materialism (Goswami 1989, 1993). Under the aegis of this philosophy, consciousness is looked upon as the ground of being in which matter exists as waves of possibility. Downward causation of the event of collapse consists of consciousness choosing the actuality from possibility. This concept is further explained in chapter 2, which discusses how consciousness creates the material world via quantum physics and the event of collapse.

Undoubtedly, the upholders of genetic determinism will point out a drawback here: Genes need not be treated as possibility waves; for them, deterministic Newtonian physics is sufficient. But this view is not justified. As the physicist W. M. Elsasser (1981, 1982) pointed out decades ago, gene mutations are quantum processes and must be treated as such, as changes in quantum waves of possibility. In macroevolution, consciousness creates new species by creatively choosing from genetic variations that exist as quantum possibilities (see chapter 11 for further details).

We will see that this theory not only explains the fossil gaps (seeing them as examples of biological creativity) but also makes room for the few observed cases of intermediates. I should also emphasize that in contrast to geographical isolation theory, the theory of creative evolution has consequences that provide independent verification of the theory (see chapters 11 and 12; also see later in this chapter).

The Nobel laureate physicist Paul Dirac once said that the solution of great problems requires the giving up of great prejudices. Darwin had to give up his prejudice for Christianity and its doctrine of biblical creationism so that he could explain the data he and his contemporaries collected. In the twentieth century, physicists had to give up the great prejudices of causal determinism and continuity in favor of quantum indeterminacy and discontinuity. Today, the twenty-first century demands an equally revolutionary change in the mind-set of biologists. They must give up the prejudices of genetic determinism and the Darwinian continuity of all biological evolution. The remainder of this chapter looks at some specific areas in which prejudices need to fall away.

The Biological Arrow of Time

In the ultimate reckoning, there is something dissatisfying, on a fundamental psychological level, about Darwinism as a theory of the evolution of life. It cannot tell us about anything meaningful about the future of our evolution. Intuitively, we recognize that human consciousness has become vastly more sophisticated as society has changed over the millennia (the subject of chapter 17), and we sense that this evolution will continue. However, both Darwinists and creationists miss the importance of this evolution of consciousness.

Darwin himself was very aware that one of the shortcomings of his theory was that it had no role for the obvious purposiveness of evolution. If biologists took this hint from their hero and followed up on this question, they would readily become aware of a crucial aspect about evolution that they have been mostly missing: the arrow of time.

In everyday life, we take for granted that we can distinguish between the past and the future. However, explaining this directionality, this "arrow of time," is actually an important problem in physics. In fact, physicists still debate about the correct dynamical explanation, but phenomenologically at least they have found an answer in the form the entropy law. Entropy is the amount of disorder of a system. Simply stated, the entropy law says that entropy (disorder) always increases, so by looking at entropy, one can distinguish past and future.

However, physicists have also been quite aware of another way to tell the past from the future. Look at the relative complexity of biological beings; look at the fossil data. Typically, old fossils show simple creatures. Newer fossils are more complex, reflecting evolution. This march of complexity is also a valid way to tell the direction of the arrow of time. In other words, in addition to an entropic arrow, there is a biological arrow: Evolution proceeds from simple to complex.

Physicists (see, for example, Davies 1988) have pointed out that Darwinism (and its offshoots) cannot be the correct theory of evolution because it has no mechanism for the biological arrow of time. According to Darwinism, evolution takes place through chance and necessity. Chance is obviously random; no preferred directionality there. Necessity, natural selection based on survivability, depends on fecundity, that is, the success of the species in making babies; complexity has no role.

The physicist Ilya Prigogine (1980) has attempted to develop a theory of evolution that includes the trend toward complexity in biological systems. The theory proposes that, with the passage of time, matter produces increasing complexity of self-organization. Although Prigogine received a Nobel prize for the initial promise of the work, the theory has not lived up to that promise in application.

One of the most important aspects of the theory presented here, creative evolution, is that it explains the biological arrow of time. The explanation is based in the purposiveness with which Darwin himself struggled. Purposiveness can be incorporated because this theory depicts evolution as evolution of consciousness (see chapter 5). Thus creative evolution, unlike Darwinism, has something to say about the future. Although Darwinism is satisfactory for explaining the slow epochs of evolution, the really interesting things happen during the fast epochs—the punctuation marks. That's when creative leaps take place, sometimes gigantic movements in consciousness. And in these movements, the purposive evolution of consciousness through complexity, consciousness gains the ability to experience itself with ever-greater sophistication.

THE OTHER, QUIET CONTROVERSY: ORGANISMIC BIOLOGY

While creationism and intelligent design theory have been attracting public controversy, another, private controversy has been raging within biology. This one pits a relatively small number of individuals, whom I call "organismic" biologists (see Strohman 1992; Ho 1993), against the mainstream upholders of strict scientific materialism. Organismic biologists emphasize the importance of accounting for the development of macroscopic traits, and eventually the development of the entire organism, through evolution (Ho and Saunders 1984).

The importance of considering the development of individual traits becomes obvious if we ask the question this way: What good is one quarter of an eye, or one tenth of a wing? A trait is useful for survival only when it is fully developed. Organismic biologists hold that the evolutionary emergence of a new trait cannot be a gradual, Darwinian, bit-by-bit process. Pieces of microscopic genetic variation that produce only a small fraction of a macroscopic trait would be eliminated by natural selection; they have no survival benefit. Consider the following little tale of partial change.

A violent criminal was taken prisoner and put under twenty-four hour watch lest he escape. The fellow had a wound in his left leg that went untreated; it developed gangrene. The prison surgeon decided the leg had to be amputated. The prisoner, being highly religious, wanted to have the leg after it was removed so he could arrange for its proper burial. All this was arranged; in the evening a friend of the prisoner's came, and the prisoner gave him the leg. The prison guard did not know about all the underlying arrangements and became suspicious.

Later, the prisoner's right leg also developed gangrene and had to be amputated. The same arrangement was made as before. The friend came back in the evening and picked up the right leg. This time the prison guard got *really* suspicious, and he blurted out, "What is going on here? Are you trying to escape?"

Like this prison guard who thought that a prisoner could escape piece by piece, leg by leg, Darwinists think that an organ can evolve bit by bit. Sometimes they argue that a quarter of an eye is not useful for seeing, but it might have been useful for some other purpose during the course of evolution. But for something as complex as the eye, which requires thousands of genetic changes, this argument would claim thousands of alternative uses for all those intermediate, developmental stages of the eye. Doesn't sound worthwhile, does it? Moreover, the issue of finding intermediate fossils for all those intermediate stages wouldn't go away.

As another example, think of the evolution of the long neck of the giraffe (Hitching 1982). According to Darwinists, this change happened through slow, accumulated variations and their selection, gradually leading to longer and longer necks that enabled giraffes to beat the competition and reach higher and higher branches of trees where the food is abundant. This scenario is too simplistic. Longer neck vertebrae require many concurrent modifications. As the vertebrae become longer, the head must become smaller, because it becomes more difficult to support the head atop a long neck. The circulatory system has to produce higher blood pressure; valves must originate to prevent overpressure when the giraffe stoops to get a drink. The lung size has to increase so the animal can breathe through a much longer pipe. Additionally, many muscles, tendons, and bones have to change harmoniously; in fact, the entire skeletal frame has to be restructured to accommodate lengthened forelegs. It goes on and on. Clearly, much more than neck-lengthening gene mutations have to be involved—and with what amazing coordination! All this through cumulative step-by-step chance and necessity? It's simply not credible.

But in creative evolution, all these changes can accumulate in quantum possibility until consciousness suddenly collapses the right set of changes into a single, new gestalt (see chapter 11 for details). As you will see, the theory of creative evolution incorporates many ideas from the organismic biologists (see chapters 11 and 12).

THE OLD CONTROVERSY:
LAMARCKISM

One scientific controversy that just won't go away is associated with the name of the biologist Jean Baptiste de Monet, Chevalier de Lamarck, whose major work predated Darwin's by about forty years. Lamarck is famous for his idea of the inheritance of acquired characteristics. Individuals of a species sometimes undergo quite universal bodily, or somatic, changes because of environmental factors. If such somatic changes could be passed on to future generations, we would have an alternative to Darwinian evolution—Lamarckism. For example, how did we humans acquire such thick skin on the soles of our feet? Lamarckism could explain it most straightforwardly: Harsh environments forced a generation of barefoot wanderers to acquire thick soles, and subsequent generations acquired this change through heredity even when the environment had returned to normal.

Unfortunately, information does not seem to flow from the soma (i.e., the body) to the genes (Weisman 1893), so biologists in general have been very unfriendly to Lamarckism. When James Watson and Francis Crick formulated molecular biology in the 1950s, they made the unfriendliness into a dogma. For Crick, it was *the* central dogma of molecular biology—information can only flow one way, from the genes to the proteins; never the other way, from proteins to genes, as Lamarckism would imply. Environment can have direct effects only on the proteins of the somatic cells, and this information cannot be communicated to the genes of the cells involved with reproduction.

Do you see the inherent bias here? The bias says that genes are the only entities that can carry hereditary information. This prejudice has run its course. In the next few chapters, I will show theoretically and empirically that new entities and organizing principles must be introduced in biology to make sense of unexplained data and theoretical paradoxes. I will show later that these new entities open alternative avenues for conveying hereditary information from generation to generation (see chapter 20). One of the successes of the theory presented here is that Lamarckism makes a long-overdue comeback.

Direct experimental verification of one of Lamarck's basic ideas has now come to fruition. Lamarck thought that an organism has within itself a way of responding actively to the environment and effecting its own evolution. The new evidence suggests that certain bacteria, when threatened with mass starvation, accelerate their own mutation rate to evolve to a new species that can survive on the available food (Cairns, Overbaugh, and Miller 1988). This behavior is called *directed mutation*. Critics of directed mutation point out that under starvation perhaps the mutation rate of all the genes is enhanced, not just the one needed for survival. But even so, the question remains: What enhances the mutation rates? The correct explanation is to see this phenomenon as direct evidence in favor of downward causation (Goswami and Todd 1997) and the causal efficacy of organisms, as also propounded by organismic biologists.

In short, the time has come to adopt a new biology, one armed with new organizing principles and with the ideas of downward causation and biological creativity. This new biology reconciles many disputes, integrates many ideas and theories, and, most importantly, satisfactorily establishes the relationship of biology and physics, as we will see in the next three chapters.

2

QUANTUM PHYSICS *and* DOWNWARD CAUSATION

Quantum objects are waves of possibility. You know the most striking property of waves; they can be at more than one place at the same time. Two people at different places but at the same distance from a sound source hear the sound wave at the same time. Quantum objects also spread like waves, and they also possess the wavelike characteristic of having more than one facet at the same time. Their only special feature is that the simultaneously occurring facets of quantum objects are *possible* facets. In one observation, one facet may show up; in another, a different facet. As for which facet is going to show up in a given experiment, only probabilities can be calculated, hence the name *wave of possibility*. A choice is needed to convert possibility to actuality. Who chooses? Because it is our observation of a quantum object that collapses the quantum wave of many possible facets into a unique event of actuality in our experience, we must say that it is our consciousness that chooses. So in quantum physics, the agent of downward causation is recognized as the consciousness of a human being (von Neumann 1955; Goswami 1989, 1993; Stapp 1993).

At first, it sounds like a statement from the old Pogo cartoons: We have looked at God, and it is us. The idea seems to be too subjective to qualify as science. But it really is not. When we examine the properties of a consciousness that can "collapse" or convert the quantum wave of possibility into an actual event without raising logical paradoxes, we find three very telling characteristics (Goswami 1989, 1993):

1. Consciousness is the ground of all being.

2. This choosing consciousness is unitive and nonlocal—that is, it communicates its choice without using signals—and is the same for all of us. In other words, the choosing consciousness is objective.

3. In the event of a quantum collapse, consciousness becomes "self-referent" in us, not only giving us the sensation of the manifest object but also the experience of a self—a subject that senses the object as separate from itself.

To put it a different way, we choose not from our ordinary ego-consciousness, but from a nonordinary state of unitive consciousness—call it *quantum consciousness*. You can easily recognize, though, if you are familiar with esoteric spiritual traditions, that this unitive character of consciousness is widely recognized as God-consciousness. Quantum physics is introducing God-consciousness as the agent of downward causation.

You now can see that the use of the phrase "downward causation" is quite justified, because this causation arises from a nonordinary God-consciousness within us that is "higher" than our ordinary "lower," individual ego-consciousness.

The same idea of a God-consciousness has found its way into transpersonal psychology (Maslow 1971; Wilber 1996) as the agent of psychological transformation and even into medicine as the agent of spontaneous "quantum healing" that occurs without any medical intervention (Chopra 1990; Goswami 2004). Incidentally, our capacities to transform and to spontaneously heal ourselves are both splendid examples of scientific evidence of the causal efficacy of human intelligence.

I will give details later, but I hope you begin to see that creative evolutionism as defined here is part of a whole new trend in science—part of a genuine paradigm shift to include downward causation (Goswami 2008). Does biology want to be left behind as the last science claiming to be based on the "single vision" of upward causation and genetic determinism? Are biologists such a conservative bunch that they won't examine the efficacy of downward causation to solve the genuine problem of the fast tempo of macroevolution? Please note that it is not an either/or choice. Both upward and downward causation are validated in this new approach. Upward causation dynamics produce the quantum possibilities in a given situation; the downward causation by quantum consciousness, God, chooses among these possibilities to collapse the possibilities into the unique actuality found in the manifest world.

CONTINUITY AND DISCONTINUITY

I must address one question before going any further. Why aren't we aware, in ordinary life, of our connection with God-consciousness? Why do we experience ourselves as individual egos instead? The answer, in brief, is conditioning. In God-consciousness, we have total freedom to choose among the possibilities that quantum dynamics offers for the states of quantum objects. Conditioning limits this freedom of choice in favor of past responses to stimuli (learning). Eventually, we become conditioned to identify with a particular pattern of habits for responding to stimuli; this identification is the ego (Mitchell and Goswami 1992; Goswami 1993; see also chapter 18).

Something else is important here. Look at figure 2. Quantum possibility waves spread in "potentia" (a name physicist Werner Heisenberg gave to the domain of quantum possibilities), not in space and time but transcending them; this part of their movement is continuous. But when consciousness chooses out of the possibilities to create actuality in space and time, the collapse of the wave of possibility is not gradual in space and time but is rather a discontinuous event for which a mathematical or mechanical description is impossible (von Neumann 1955). As the physicist Niels Bohr first recognized, this discontinuity is part and parcel of quantum physics. Bohr called this discontinuity a "quantum leap."

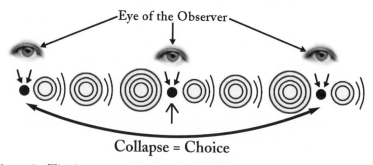

Figure 2. The (continuous) spread and (discontinuous) collapse (due to conscious observation and choice) of quantum possibility waves.

However, with conditioning, the discontinuity of quantum collapse is obscured. We seemingly experience our ego consciousness as a continuous stream.

In other words, when we are in God-consciousness, our choice from quantum possibilities is creative and discontinuous — it is a quantum leap. In ego-consciousness, the choice is restricted to conditioned alternatives: It emanates from the past and continuity seems to reign. Some freedom exists for choice, but it is not creative freedom. We cross the border when we say "no" to conditioning, thus becoming open to creative freedom.

If you are having difficulty visualizing a creative quantum leap, an example from the electronic movement in atoms à la Bohr may help. Atoms emit light when an electron jumps from a "high" orbit to a lower orbit. But the jump is discrete, quantum; the electron is here in one orbit and then there in the other, never physically going through the intervening space between the two orbits. Creativity can have a similar feel.

Creativity researchers have shown that the creative process has four stages: preparation, incubation or relaxation, sudden insight, and manifestation. Preparation is work, and so is manifestation—manifesting the product of a creative insight. The necessity of work for creativity is easy to understand. But it is well known to creativity researchers that if we intermittently relax (just sit as a bird sits on its egg in incubation) while actively researching a problem, creative quantum leaps of insight are much facilitated (Goswami 1999). But why? Psychologists call this stage of the creative process *unconscious processing*. More mystery is added.

Quantum thinking gives us the answer and solves the mystery. Realize that before choice, before collapse, all quantum objects, being waves of possibility, spread as a water wave spreads after you throw a pebble in a pond. The process is a little more subtle for quantum objects, because their waves spread in possibility, becoming bigger and bigger conglomerates of possibilities from which consciousness may choose. Why call the processing of possibilities "unconscious"? Because without collapse, the subject-object separateness required for awareness is absent (see chapter 5). The Freudian (and somewhat vague) concept of unconscious really means unaware; as you can see, the concept can be defined very clearly in quantum physics when explained within the primacy of consciousness. (For further details, see chapter 15.)

In this way, through the phenomenon of creativity, both the continuous and the discontinuous aspects of the quantum movement of possibility waves are used.

It is of utmost importance that biologists catch on to the new message of quantum physics: Movements in the world can be both continuous and discontinuous. In classical physics, there is only continuity, and because, by and large, the quantum message has not penetrated biologists' psyches, they still hold on to the continuous, Darwinian scenario of evolution.

CONSCIOUSNESS WITHIN BIOLOGY OR BIOLOGY WITHIN CONSCIOUSNESS?

Let's face it. The creationism-evolutionism controversy has been so politicized that most biologists must feel extremely uncomfortable entertaining the idea that God can have anything to do with biology. When biologists realize that quantum physics views God as "objectively defined cosmic consciousness," they should feel reassured, or so I hope.

So let's examine in some detail the logic behind the assertion above. First let's discuss how the idea that consciousness is the ground of being is forced upon us by quantum physics. Take the idea that conscious choice affects the quantum possibility wave of an object by collapsing it into an actual event of our experience, into a "particle," so to speak. This idea seems dualistic at first. Why? Because consciousness has to be nonmaterial to effect collapse. To see this, suppose, as materialist biologists believe, that consciousness is a brain epiphenomenon. But undoubtedly the brain is a conglomerate of elementary particles, quantum possibilities, so it must itself also consist of quantum possibilities. Ditto for any epiphenomenon associated with it.

Now do you see why consciousness, to effect collapse, must be nonmaterial? A material consciousness arising in the brain is only a possibility wave. A possibility wave acting on a possibility wave just makes a bigger possibility wave. No actuality ever comes out of such an interaction (von Neumann 1955).

But if consciousness is nonmaterial, then we still have the problem of dualism: How does a nonmaterial consciousness interact with a material possibility wave? Scientific materialism avoids dualism by adopting a viewpoint called *material monism*—everything is made of one thing: matter. Or to put it differently, matter is the ground of all being. But the dualism is also resolved if we posit a different kind of monism. Suppose we posit instead that consciousness is the ground of all being and everything is made of consciousness. Then matter consists of possibilities of consciousness itself. Collapse of a quantum possibility wave of matter into the material actuality that we experience consists of consciousness choosing nonlocally out of its own possibilities, and dualism does not arise.

A look at gestalt pictures (fig. 3) may help here. Look at the first picture (fig. 3a) and you will see either a vase or two faces looking at each other. In the same way, a look at figure 3b will reveal either a young woman or an old woman (the figure is called "The Wife and the Mother-in-law"). Here is the important point. If you are seeing the vase (or the young woman), you can shift to seeing the faces (or the old woman) simply by shifting your perspective slightly. Try it! But are you doing anything to the picture to accomplish this amazing feat? No, because both meanings are

Figure 3. Gestalt pictures. The meaning changes without our doing anything to the pictures. (a) The vase and two faces; (b) the wife and the mother-in-law.

already in your consciousness. You are just choosing by recognizing one or the other of the two meanings. Consciousness choosing one facet out of the many-faceted possibility wave works something like this. There is no doing involved, no dualistic interaction, no signals. Just recognition and choice.

The lesson of all this is that the paradigm must shift. As long as biology treats consciousness as an epiphenomenon of the brain, it is not compatible with quantum physics. To bring biology into the realm of quantum physics, we must reformulate biology and place it within a new metaphysics, the metaphysics of the primacy of consciousness.

The Objectivity of the Choosing Consciousness

Can we do science based on the primacy of consciousness? That is, can science be based on the idea that consciousness is the ground of all being and that all things of experience are chosen from quantum possibilities of consciousness? Where is the necessary objectivity? Now the quintessential proof: I'll show why the choosing consciousness of quantum physics has to be an objective cosmic consciousness.

You may have heard of the famous paradox of Schrödinger's cat. In this "thought experiment," a cat is left in an opaque cage with a door in the company of a single radioactive atom (with a half-life of one hour, which means that the probability is fifty-fifty that the atom will decay within the hour). The scheme is diabolical. If the atom decays, then a Geiger counter detects the radiation from the decay. The ticking of the counter triggers a hammer to break a poison bottle, releasing cyanide that kills the cat. If the atom does not decay, none of the sequential things happens and the cat lives. Because the decay is probabilistic, a quantum process, it follows that at the end of the hour the cat is a wave of possibility, having two equally weighted facets, dead and alive. The cat is literally half-dead and half-alive!

This half-and-half state is an example of quantum weirdness, but it doesn't constitute the paradox (we'll get to that in a second). This state follows strictly from quantum mathematics. Because every consequence of quantum mathematics has been experimentally verified with uncanny accuracy, we must assume that this explanation accurately describes the cat's situation when we are not observing it.

Of course, when we observe it, we find the cat ether alive or dead, not half-alive and half-dead. This is a paradox from the point of view of scientific materialism, because all objects are quantum possibilities; an observer's brain is no exception. So why should an observer's looking—an operation of a quantum possibility wave, namely a brain—be tantamount to a choice between the cat's life and death?

However, if you have grasped the discussion above, you will see that if we adopt the metaphysical position that consciousness is the ground of all being, then indeed conscious looking is tantamount to choice and produces collapse.

The famous physicist John von Neumann (1955), who is also famous in biology for his work on a concept known as *cellular automata*, proposed conscious choosing as the resolution for the plight of Schrödinger's cat, but the solution didn't hold for long. Soon after, the Nobel laureate physicist Eugene Wigner raised a further paradox. This is called the paradox of Wigner's friend.

Suppose Wigner has a friend in his lab; they visit while the cat waits in its cage as a possibility wave suspended between life and death. After the hour, out of politeness Wigner invites his friend to look over his shoulder as Wigner opens the cage. Because both are looking at once, whose looking is going to collapse the state of the cat? In other words, who gets to choose?

The paradox becomes acute if we posit that Wigner is a cat fancier: He would choose the cat to be alive. But his friend despises cats. No doubt his choice would be thumbs down: Let the cat die! So again, whose choice counts? Does the cat live or die?

One solution is to posit that only Wigner is conscious, that his friend is a figment of Wigner's imagination. This philosophy is called *solipsism*, and surely we all feel that way at times. But such a philosophy would not be scientifically useful, although, ironically, materialist metaphysics cannot refute it either.

When you think about it, solipsism is not really a solution to the paradox; it just shifts the question from "who gets to choose?" to "who gets to be the solipsistic head honcho?" The real solution, when it emerged, created a stir. Three physicists, Ludwig Bass (1971) in Australia, myself (Goswami 1989, 1993) at the University of Oregon, and Casey Blood (1993, 2001) at Rutgers University, all working independently, discovered the same resolution. The choosing consciousness is cosmic and unitive, transcending individuality.

In other words, we don't choose the cat's fate from our ordinary individual consciousness, what psychologists call *ego*. We choose from a nonordinary state of consciousness in which our ordinary individuality gives way to a cosmic oneness—an objective consciousness beyond our individual conscious identity.

Thus not only is the paradox resolved, but our effort to locate science within consciousness is back on track, because consciousness is objective after all. For a large number of observations, quantum consciousness would choose with apparent objectivity, as probability calculus dictates. The suspicion that subjective quirkiness plays a role in the affairs of the world is gone. And yet downward causation is still choice. That is, consciousness does not necessarily follow the dictum of probability in any individual event of choice. In any individual event, there is always room for creativity.

As Erwin Schrödinger (1944) himself wrote in the appendix of his seminal book, *What is Life?* "Consciousness is a singular for which there is no plural." In its base nature, consciousness is cosmic, the one and only. Our ego is an illusory, separate individuality that arises because of the identification of consciousness with the brain and subsequent conditioning (Goswami 1993).

This oneness of our consciousness, that consciousness is nonlocal, was the first prediction of the theory presented here. This prediction has been experimentally verified and replicated (Grinberg-Zylerbaum et al. 1994; Sabel, Clarke, and Fenwick 2001; Wackermann et al. 2003; Standish et al. 2004). In the original Grinberg experiment, two subjects meditate together for twenty minutes with the intention of nonlocal connection, an intention they meditatively maintain even when they are separated after the twenty minutes and put in individual Faraday cages (enclosures impervious to electromagnetic signals). The subjects are connected to separate electroencephalograph (EEG) machines that record their brain waves. One of the subjects is shown a series of light flashes, which induce electrical activity in the brain that is recorded by the EEG. A computer eliminates the noise from the EEG signal and extracts a signal called an *evoked potential* from the recording. Surprisingly, a similar potential shows up in the EEG recordings of the correlated partner who, however, has not seen any light flash. If electrical potential can be transferred from one brain to another without any signal, a nonlocal connection must exist between them. This nonlocal connection is their unitive quantum consciousness. Incidentally, these experiments also rule out solipsism as a viable solution of the paradox of Wigner's friend.

Quantum Creativity and Entropy

Let me emphasize again that the unity of consciousness solution of the paradox of Wigner's friend, given above, allows creativity for a single object in a single event. For many objects and many events, the entropy law—that disorder always increases (or at best remains the same)—prevails. It follows from quantum physics-within-consciousness when we realize that in the absence of a creative urgency, consciousness collapses the possibility waves in accordance with the needs of the statistical average, that is, in accordance with the probability distribution. Therefore, I'd like to propose a restatement of the entropy law:

> In the absence of quantum creativity, that is, creative collapse, things tend toward maximizing probability, which is the same as maximizing entropy.

This reformulation opens up the possibility that, in the presence of quantum creativity, the door is open for things to change in the direction of increasing order and complexity.

SOME CONSEQUENCES

Coming back to Schrödinger's cat, in my quantum physics classes I have taught the Schrödinger's cat paradox many times and inevitably a cat lover gets irritated and asks, "Why can't the cat decide whether she is alive or dead? Is the cat conscious or not?" When Schrödinger decided to make the paradoxical aspect of the quantum observer effect more spectacular and personal by using a cat, I think he may have been prejudiced to believe that cats are no different from inanimate, insentient matter, that cats are machines (a belief held by Christianity and shared by the philosopher Descartes). In this view, cats don't have the ability to collapse their quantum wave, and so the paradox of the observer effect can be illustrated using the cat.

The problem is that present-day materialist biology shares Schrödinger's prejudice: Because life is considered to be nothing but the physics and chemistry of macromolecules, it could as well be a human in the Schrödinger cage. As already noted, this stance is not compatible with quantum physics, the best-ever theory of physics and one that has been verified by myriad experiments with unprecedented precision.

In other words, quantum physics demands that biologists give up their materialist prejudice and base biology on the metaphysics of the primacy of consciousness. One of the most important rewards of such a change of paradigm is no less an accomplishment than being able, for the first time in biology, to clearly distinguish not only between the conscious and the unconscious, but also between life and nonlife (see chapters 7 and 8). So, yes, not only we humans but cats and lizards and even one-celled organisms can collapse possibility waves into actual events of experience. Incidentally, this distinction will make use of the third characteristic of consciousness introduced above, the characteristic of self-reference.

A few words for the God aficionado who may feel left out when God is replaced by cosmic quantum consciousness. Behold! This way of looking at God is common to all spiritual traditions at their esoteric core. It just gets lost in popular translation. Jesus knew about it, as did Buddha, Lao Tzu, and the *rishis* of the Hindu Upanishads. The Sufi master Ibn al-'Arabī knew about it too. Quantum physics is just enabling us to rediscover truths our forebears discovered in their own ways. In the process, quantum physics also reassures us that there is nothing unscientific about God, that God and the idea of God's creativity can be incorporated in an objective science, in particular, in a new biology. Both we and biology will be better for it.

And where does creationism fit? The followers of creationism, if they base their theories not on the Old Testament but on the New Testament, would readily agree with the quantum view that God-consciousness is actively involved with creation everywhere and at all times, not just once six thousand years ago. The truth is, evolution is the wrong target for God aficionados to be protesting. The real target of alarm should have been the exclusive doctrine of upward causation, scientific materialism, and Darwinism.

Anyhow, all this is moot once you see the cogency of a new biology within consciousness, when God-as-quantum consciousness is recognized as the organizing principle of creative evolution in biology.

THE NEW BIOLOGY

All the political wrangling is an unfortunate distraction from the real issue in biology: the need to find new organizing principles. Note the plural: Quantum downward causation is not the only new organizing principle needed. When we delve deep into the relation of physics and biology, very soon we discover a vast abyss between the two that biologists hardly acknowledge, though scientists outside biology are painfully aware of it. (This abyss is the subject of chapter 3.) We further discover that the abyss between physics and biology can only be bridged if we introduce new organizing principles that govern the movement of extraphysical components of consciousness that we experience inside of us—feeling, thinking, and intuition. This move also aligns with intelligent design theory. The point is that a designer needs not only raw material, but also blueprints for making the design. The designer is God, cosmic quantum consciousness. The raw material is matter, the biological macromolecules and all their attendant processes. Biologists have established a sophisticated science of the raw material: the field of molecular biology. But what acts as the blueprints? This question is the subject of chapter 4.

3

CONNECTING BIOLOGY

to PHYSICS

*I*f we want to base biology on a materialist foundation, as most biologists implicitly do, then the parent-offspring relationship of physics and biology must be clearly demonstrable. In other words, biology should reduce to physics. The question is, can biological phenomena be regarded as phenomena of physics, with the apparent differences between physical and biological phenomena arising from the sheer complexity of the biological situation? We can think about this kind of reduction in several ways. First is the question of components, entities, or processes. For example, the components for physics (and its inanimate objects) are molecules, atoms, and elementary particles; are the components the same for the animate kingdom of biology? We can also make the reduction a question of theory: Can we theorize about life and living processes with the theories of inanimate matter that constitute physics? Yet another track is to ask this: Can we use the same methodology or research strategy in biology as we do in physics?

Most biologists, even today, think that the first reductionist question, namely, what components constitute life, was resolved once and for all when the philosophy of vitalism (the idea of a living being having a dualistic life-giving vital body in addition to its physical body) was discredited with the rise of molecular biology. To these mainstream biologists, life is an emergent property of the complex molecules of the cell interacting in complex ways. Just as the liver secretes bile, a cell secretes life.

As to the question of theory, many biologists would flatly declare that ultimately biological theories must be reducible to physics. But "ultimately" is far away, so at this stage biology might as well have its own theories.

As to methodology, biology is rapidly becoming a laboratory science of molecular biology, much to the dismay of field biologists such as ecologists.

Can biology be reduced to physics? Francis Crick represents the sentiments of most biologists when he says, "The ultimate aim of the modern movement in biology is in fact to explain all biology in terms of physics and chemistry."

The comedian Woody Allen used to do a comedy routine (long ago, before he became a famous moviemaker) in which he goes to England and buys a Rolls Royce. He wants to bring it back to America but doesn't want to pay import duty. What to do! Ingeniously, he dismantles the car and packs the parts in a whole bunch of suitcases and manages to return to the States without paying import duty. But afterwards, when he tries to reassemble the Rolls, to his dismay he can't. He makes sewing machines, a tank, even an airplane, but never the original car. The complexity was too much for him.

Some scientists feel that the problem with cavalier reductionism in biology is the sheer complexity of biological forms. To solve that problem, they try to develop a special science of complexity theory to tackle biological reductionism. This is also a myopic view, because it fails to recognize the consciousness revolution that quantum physics has brought to physics itself.

In physics though, although quantum physics requires consciousness for its interpretation, physics in the main can be carried out without invoking consciousness. This seeming paradox can exist because in physics and chemistry we are always dealing with zillions of molecules. Quantum waves of possibility have associated probabilities that can be calculated using quantum mathematics. When we are dealing with large numbers of quantum objects, this probability calculus allows us to derive extremely accurate predictions that can be verified by laboratory experiments. In most of physics, downward causation is only the metaphysical backdrop.

But in biology the situation is quite different. We have already seen in the last two chapters that at least one major problem of biology, namely the fast tempo of macroevolution, demands a new organizing principle—downward causation—operating explicitly, alongside explicit upward causation. Does this claim suggest that we are proving that, in a consciousness-based science, biology can be reduced to physics, with the only difference being that in biology the role of downward causation is often explicit, not implicit? The answer is no: Biology needs still other organizing principles. I will take up this subject in chapter 4, but first I want to further clarify the nature of what biology misses and show where creativity fits in the new biology.

MORE ON THE INCOMPLETENESS OF BIOLOGY

Biologists need to come to terms with the incompleteness of biology. Part of the problem is the bigness of the biology enterprise, and part is the bigness of science as a whole. Science is so huge today that it has to be done piecemeal. What one hand does, the other hand doesn't know about.

Biologists treat the problem of the origin of life separately from the problem of the origin of the universe and, of course, from the quantum measurement problem. The solution of the nature and origin of life, we will see in chapters 7 and 8, lies in an integral view.

Even biology is so huge that it is done piecemeal. The evolutionary biologist is under the impression that dualistic talk has been eliminated in biology, without realizing that mind-body dualism is quite alive in the neurophysiologists' theories about how the brain makes representations of an object of perception (see chapter 16).

Similarly, origin-of-life biologists talk about "meaningful information" (as in DNA) but ignore the need to show how it is possible for a machine to process meaning. The latter issue is raised only in the context of the mind-body problem.

Before molecular biology came along, Western science endorsed vitalism, the idea that a "vital body" infused life into matter. When biologists claim to have exorcised vitalism with molecular biology, are they aware that the Eastern medical systems of acupuncture and *ayurveda*, and the Western system called *homeopathy*, all crucially depend on the concept of vital energy? Are they aware of the causal efficacy of these healing traditions— efficacy that cannot be explained away with materialist models? Do today's biologists know that *élan vital*, a term coined by philosopher Henri Bergson early in the twentieth century, was a popular notion in biology for a time? Can they explain feeling, which is our experience of vital energy? Are they aware of recent work that connects biological forms and feeling, chakra medicine and chakra psychology (Goswami 2004)?

Many biologists believe that molecular biology has sufficient explanatory power for everything connected with the life of a cell. However, another group of highly dedicated scientists called *complexity theorists*, a group that includes more than a few biologists, are convinced that life arises from self-organization that goes beyond molecular biology (Kaufman 2002).

A few biological theorists are even aware that a living cell has the capacity to cognize, that is, to be aware of a specific piece of information, which implies a cognizing self (Maturana 1970; Lipton 2005). But how does this self, a subject, arise when the starting point is objects, the macromolecules of the cell? Are cell biologists aware of the problem of explaining subjective qualia (felt experience of qualities) in a neurophysiological theory of perception?

Are biologists aware of depth psychology (of Freudian origin) or transpersonal psychology? These viewpoints define successful sciences of psychology that help people heal themselves and reorganize their lives, but both fields make assumptions about consciousness that are radically different from that of the biologist. Consciousness in depth psychology and transpersonal psychology is the reservoir of creativity; the adaptive consciousness of the evolutionary biologist is a causal flop. But if a materialist approach is taken to be correct, shouldn't psychology be reducible to biology? Should we not worry about reconciling such contradictory points of view about the nature of consciousness?

The truth is that molecular biology of a cell explains neither an experiencing self nor feelings. Could it be that the necessary organizing principles are missing? Could consciousness explain the experience of the self? Could the vital body explain the experience of feeling?

The unfortunate truth is that when biologists are shoved against the wall, almost all resort to evolutionary adaptation as the solution. Consciousness? Of course it is the product of evolutionary adaptation, the biologists insist, forgetting conveniently the problem of the experiencing self. Let the neurophysiologists worry about the hard problems!

This won't do any longer. Biology has to be reconciled with quantum physics, including its measurement problem. Biology also has to be reconciled with medicine and psychology. Ultimately, biology also has to reconcile itself with questions of religious concern: What happens after death? Should we live our lives in accordance with ethics and values? There are the philosophical questions, too: How deep is our ecology? Whence comes our sense of aesthetics? Turning all such questions into a question of evolutionary adaptation is akin to what religions did with their God hypothesis: If I can't explain it, then surely it's the work of God.

Biology is a science, and what is science but a bunch of laws? Biologists believe in laws of biology, even though they believe that these laws are derivable from the underlying biochemistry or eventually from physics. Even then they are haunted by the question: Where do these laws come from?

The biologist Charles Birch (1999) points out that there are two ways to think of the relationship of life and matter: Either you think that life is matter-like, or you think that matter is lifelike. The great philosopher Bertrand Russell chose the first way of thinking; Alfred Whitehead, another great mind of the twentieth century, chose the second way.

Biologists almost universally sided with Russell, creating a biology that is fundamentally incomplete. If life is matter-like, then life, like bulk matter, becomes virtually deterministic. Life is also limited like matter is: unable to process meaning, lacking creativity, and lacking subjective consciousness.

BIOSYSTEMS AND CREATIVITY

What happens if we consider the Whitehead alternative—that matter is lifelike? In Whitehead's philosophy, "the primary meaning of life is the origination of conceptual novelty—novelty of appetition." Birch explicates:

> Novelty is not mere change. The ever-present entropic tendency to decay is change. This tendency can be resisted in two ways. One way is by the very stable structure of many things

such as rocks. These endure by countless repetition of unchanging patterns. The second way it can be overcome locally is by creative novelty which rises above external determination to be alive. Hence life is directed against the repetitious mechanisms of the universe. There is an urge in life to meet life's as yet unrealized possibilities. (121)

In this philosophy, the difference between "nonliving" and "living" is creativity. Quantum physics has come a long way to make that contention much more than philosophy. Can matter be lifelike? It surely can: When matter and life both are possibilities of consciousness, then matter is lifelike. And then life can be creative and fulfill its urge to manifest its unrealized possibilities. This is what creative evolution is all about.

I hope that very few biologists in their heart would deny creativity at the human level, including the causal efficacy of creativity. After all, who can ignore the causal impact of Watson and Crick's discovery of the DNA structure on the biological community? But then biologists such as Dawkins would say, well, maybe creativity arises because of evolutionary adaptation.

You cannot have it both ways. If you want biology to be governed by deterministic laws, then matter cannot be creative, ever, period. If you are willing to allow quantum uncertainty to govern biological beings, then quantum creativity is present to begin with; you don't need to invoke evolutionary adaptation.

The expression of creativity in the processes of life, including evolution, is the challenge for consciousness. But the creativity of consciousness operates within laws, so we have the science of biology. However, consciousness does not just "breathe" life into matter. Instead, consciousness uses intermediates—blueprints—and then makes representations of these blueprints onto matter, invoking creative quantum collapse. In this way, living things have both determinism and quantum creativity. It's a confusing but extremely effective way of creating complexity.

Life has deterministic tendencies, no question about it, and biologists have done a very good job of codifying them. But Whitehead is also right. To distinguish life from matter is to recognize life's novelty. This distinction is so self-evident even nonspecialists are aware of it; why should biologists deny it? The challenge is this: Can biologists acknowledge that matter is lifelike and still be scientifically credible? They can. This book shows how.

THE QUESTION OF CREDIBILITY

In attempting to redirect biology toward a more inclusive definition of life, we must also heed the genuine concerns of biologists. When vitalism was discredited, the revolt did not arise from prejudice but from a legitimate concern about dualism in general (similar to the revolt against a dualistic God separate from the world or a dualistic mind apart from body). The debate is not over whether reality is monistic or dualistic; science demands monism. Here is the true question for debate: Is monistic reality, in its "suchness" as a Buddhist might say, matter, or is it consciousness? In the last two chapters I showed that to make biology compatible with quantum physics and with all the experimental data of evolution we must reformulate biology within consciousness. In the next chapter, I show that consciousness as a basis for all science, including biology, has added advantages. It allows biology to incorporate other organizing principles that are required to make biology complete. Additionally, it reconciles biology with its offspring sciences and even with religion.

A second legitimate concern of biologists is that all scientific organizing principles be as concrete as possible, not woolly, not vague. A third concern is important for all sciences. Ultimately, it is a question of experimental data. Which basis, material or consciousness, will serve the data better? Rest assured, the proposed new science addresses these last two concerns as well.

4

*B*IOLOGY *within*
CONSCIOUSNESS *and*
*O*THER *O*RGANIZING *P*RINCIPLES

*I*n this chapter I argue that there are important roles in biology not only for consciousness and downward causation, but also for other entities and organizing principles. We begin to suspect the existence of these entities and organizing principles when we realize that biological forms—cells, tissues, and organs—and indeed, the whole organism itself, have purpose. They all perform specific purposive functions. Materialist metaphysics allows only for cause, not purpose. So how can we justify the purposiveness of biological form? The biologists' answer is that purposive functions give organisms a survival advantage. But this argument has a fatal paradox built in. To see it, first consider the idea of survival.

The Darwinian theory of evolution is based on natural selection: Nature selects those organisms that are fittest to survive. In the materialist view, an organism is just a bundle of molecules that are completely specified by their physical and chemical properties. Nowhere among these properties will you find a property called *survivability.* No piece of inanimate matter has ever attempted to survive or in any way tried to maintain its integrity under any circumstances. But living bodies do exhibit a property called *survivability.*

Now the paradox. A Darwinist would say that the survivability of the living form comes from evolutionary adaptation via natural selection. But natural selection itself depends on survival of the fittest. See the circularity of the argument? Survival depends on evolution, but evolution depends on survival!

A paradox is a sure-fire sign that the basic assumptions of the paradigm are incomplete or inconsistent; they need a reexamination. Aren't biologists missing something by objectifying subjective phenomena? The philosopher Robert Efron (1968) thinks so.

> The reductionist attacks the definition and usage of every word, which has historically referred to an action of a living entity: "memory," "reflex," "free will," "cognition," and so forth. He then redefines the same word so that it will be applicable to an action of an inanimate entity. By using this epistemological technique he deludes himself into thinking that inanimate entities have the same properties found in living organisms, that a common denominator has been found, and that the problem of reduction has been "solved." The solution is primitive animism expressed in scientific jargon.

What the reductionists miss by these tricks is "living." Their tricks take the living out of the life they try to explain!

No, biology cannot be reduced to physics, not even to physics within consciousness. In biology, not only do we need nonmaterial consciousness as an organizing principle for creative evolution, but we also need other aspects of living, along with their corresponding organizing principles, nonmaterial all, for explaining the purposive behavior of biological form, behavior that we normally identify with living.

Law-like and Program-like Behavior

Whereas physical objects are purely driven by cause, or causal laws, the behavior of biological beings is guided not only by cause but by purpose. Put another way, whereas physical objects are purely law-like, biological beings are both law-like and program-like. Program-like behavior refers to behavior that follows logical step-by-step instructions having a purpose (Goswami 1994).

Think of biological beings then in analogy with computers. It is possible to think about computers in two complementary ways: from a hardware point of view in which electrons act on electrons and from a software point of view in which symbols act on symbols via programs. Similarly, in biological organisms, at the law-like physical level (hardware), molecules act on molecules. On the program-like software level, programmed forms act on programmed forms carrying out purposive functions.

Of course, no one should claim that the software's program-like behavior should follow from the hardware's law-like behavior. There goes naïve reductionism: We can never understand the program-like behavior of living form by starting from the law-like behavior of the nonliving substrata, from physics and chemistry

What about genetic determinism, though? Can we understand all the program-like behavior of biological beings from the genes alone?

For computers, we know that we, the programmers, use the physical hardware (through downward causation) to perform software functions that are purposive and meaningful. The purpose and meaning of the programs running the symbols come from us. Similarly in biology, can there be an intelligent design of purposive traits without an intelligent designer? The conventional biologists try to deny the role of consciousness/God/intelligent designer by asserting the operation of genetic determinism and selective advantage. In this claim, Darwinian adaptation programs the genes to produce purposive traits. But as I have argued above, this position cannot be justified because of the logical circularity in the argument.

Some theorists try to eliminate the programmer by introducing an idea called *boundary conditions*. A cardinal example is the operation of a laser. The laser beam is generated in a chamber that has mirrors at each end that reflect and re-reflect photons, the quanta or irreducible units of light. The mirrors constrain the photons' freedom of movement inside the laser, forcing them to behave in a cohesive, unidirectional pattern that is the unique signature of laser light. This pattern is an example of downward causation imposed by boundary conditions.

Unfortunately for materialists, it is not at all clear whether boundary conditions can impose downward causation to such a degree that the law-like behavior of nonliving molecules changes into programmed behavior of biological form. For example, not an iota of evidence exists to suggest that the macromolecules of nucleic acid that make up DNA develop their survival necessity because they are confined by a wall in a living cell.

Complexity theorists approach things a little differently. They try to simulate life from little bits of programmed material called *cellular automata*. However, their work proves only that complicated, program-like behavior can arise from relatively simple programmed behavior. In the same vein, all genetic engineering shows is that we can take bits and pieces of programmed living matter from nature and manipulate them to our advantage. In our laboratories, we have never been able to convert law-like behavior into programmed behavior without a programmer, without our conscious interference. No doubt law-like matter is programmable, but can it be propelled from law-like behavior to program like-behavior without any help from consciousness? The answer is a resounding no.

Consciousness, that is, a programmer using downward causation, is needed to produce programmed biological form. God makes man in His own image. To figure out whether consciousness uses any other organizing principle in producing living forms, we can study what we do to write a program for a computer. We use mental imagination, patterns of meaning really, and map them as software into the computer hardware. In other words, we use a blueprint, in this case, the meanings that arise in our minds.

The program-like behavior is built into biological form making, a process that starts with a one-celled zygote. Technically, the form-making process is called *morphogenesis*; *morph* means "form" in Latin and *genesis*, of course, means "creation." The blueprints of biological form making are called *morphogenetic fields*, the fields that help program biological form.

The morphogenetic fields help to provide the cells with the programs of cell differentiation crucial to developing all the tissues required for life. How is it that the liver cell functions so differently from the brain cell when they were both made through cell division from the same original one-celled zygote? The answer is that the liver cell is differentiated from the brain cell by programs that enable different sets of genes to make different sets of functioning proteins in these two organs.

But where do these fields reside? In the genes? In the epigenetic material (Ho and Saunders 1984), that is, the cytoplasm surrounding the cell nucleus? In view of the discovery of regulator genes—genes that regulate the behavior of other genes that code for protein making—many biologists (for example, Carroll 2005) have been enthusiastic about a genetic explanation of these programs of morphogenesis. Some biologists theorize that morphogenetic programs originate in portions of DNA, called "junk DNA," that have no obvious functions. There is, however, a fundamental difficulty with thinking in this fashion. The biologist Richard Lewontin (2000) elaborates:

> ... the processes of differentiation of an unspecified cell into a
> mature specialized form are not understood. ... The problem
> of cell differentiation, division, and movement cannot be solved

without information about the spatial distribution of molecules within cells and about how the state of the cell is influenced by neighboring cells and the surrounding environment. We need to return to the old problem of "positional information." It is all very well to say that certain genes come to be transcribed in certain cells under the influence of the transcription of certain other genes, but the real question of generation of form is how the cell "knows" where it is in the embryo. (117–18)

"Positional information" is the idea that the present position and activity of a cell provide most of the information governing the future changes in the dynamics of the cell. Developmental biologists use metaphors such as "fields" and "gradients" in explaining how a cell knows how to get somewhere. Unfortunately, how the cell "knows" cannot be understood in terms of local transfer of information alone. The point Lewontin seems to make is that the dynamics of morphogenesis has to be nonlocal somehow. But all material things interact via local signals! See the predicament?

In 1981, in a book called *A New Science of Life*, a maverick biologist named Rupert Sheldrake published an unthinkable, heretical thought: The morphogenetic fields, the source of the programs that biological forms obey, are extraphysical, non-material; they reside outside the material world. When epigenetic morphogenetic fields (morphogenetic fields residing outside the genes) were proposed (Waddington 1957), the idea was considered radical by some biologists. Just imagine, then, the reaction to the idea of nonmaterial morphogenetic fields! Indeed, in the pages of the scientific journal *Nature*, two authors declared that Sheldrake's book should be burnt.

We now have to ask, How do the nonphysical morphogenetic fields interact with the physical matter? Things in the world interact by exchanging energy via signals going from one body to the other through space, taking a finite time, but the energy of the physical universe in total always remains the same. If the physical world interacted with the nonphysical world of Sheldrake's morphogenetic fields, wouldn't some energy from the physical universe occasionally leak into the world of the morphogenetic fields? Experiments demonstrate conclusively that this never happens— the energy of the physical universe is always a constant. Sheldrake's (1981) answer was again, to put it mildly, radical: The interaction of the morphogenetic fields with physical matter is a resonance of sorts. It is nonlocal, requiring no exchange of signals through space. Such nonlocal interactions are instantaneous.

Sheldrake, being an experimental scientist, concentrated on various experimental demonstrations of nonlocality of his morphogenetic fields. Although some of his experiments succeeded, others were ambiguous, so the jury is still out on this one (see Durr and Gottwald 1997).

Fortunately for Sheldrake, quantum physics gave unexpected support for his nonlocality hypothesis in 1982 when physicist Alain Aspect and his collaborators (Aspect, Dalibard, and Roger 1982) demonstrated experimentally that quantum objects, when properly "correlated," do communicate nonlocally. In Aspect's experiment, a laser-irradiated atom emitted two photons (quanta of light), one going one way, the other moving opposite. Strangely, whatever the experimenters did to detect the one photon, the effect was communicated to the other photon without any exchange of any signals. Quantum nonlocality, yes! Sheldrake, maybe.

The adherent of conventional biology can raise one serious objection to Sheldrake's hypothesis. The scenario of interaction or communication between something material and something nonmaterial smacks of what philosophers called *dualism*, already mentioned in chapters 1 and 2. This philosophy was much touted by the seventeenth-century philosopher René Descartes and has been much discredited ever since. Two things made of different substances (such as the material macromolecules of biological cells and the nonmaterial morphogenetic fields) just cannot interact, correlate, or communicate without a mediator. "Show us a nonlocal mediator," became the philosophical demand on Sheldrake's theory.

Interestingly, quantum physics, when interpreted within the primacy of consciousness as discussed in chapter 2, solves the problem of a nonlocal mediator.

QUANTUM PHYSICS, CONSCIOUSNESS, AND POSSIBILITIES

I have spoken already of matter as quantum waves of possibility. These waves are not separate from consciousness; they are possibilities of consciousness itself, because consciousness is the ground of all being and because everything, including matter, is made of consciousness.

This statement may sound counterintuitive, but quantum physics demands this stretch of your imagination. As the physicist Casey Blood (1993, 2001) insists, quantum objects of matter are mathematical entities, those waves of possibility; their material nature—hardness, structure, texture, and all that—emerges only upon the interaction of consciousness with those possibility waves, when consciousness collapses the possibilities by choice, by observation.

Once you are able to visualize matter as quantum possibilities of consciousness without feeling a conceptual quandary, ask yourself this: Could consciousness have other kinds of possibilities? Sure, why not? Suppose the morphogenetic fields are also quantum possibilities of consciousness, though nonmaterial, to be sure. In an event of collapse, consciousness collapses not only the material waves of possibilities, but also the waves of possibility corresponding to the morphogenetic fields. In the process, consciousness makes a representation of the morphogenetic fields in matter; these representations are the biological programmed forms. Dualism is absent from this way of looking at biological form making.

The morphogenetic fields do not guide physical form by direct interaction; instead, consciousness uses the morphogenetic fields as blueprints to direct the choice of a particular form from among the myriad quantum possibilities of matter. The physical body is thus a representation of the morphogenetic blueprints. These representations are the purposive programmed forms of biological bodies.

Biological forms, thus, are representations of vital blueprints of biological functions. Once again we can make the analogy to a computer. A computer has hardware; our physical world is akin to the computer hardware. A computer also has software; the programmed forms of biosystems are akin to computer software. For our silicon computer, groups of software symbols represent our mental meaning. In the case of living beings, groups of cells—the organs—represent the morphogenetic blueprints for the purpose of carrying out biological functions, functions that include maintenance, survival, and reproduction.

Can the biological programs come about from the law-like behavior of underlying molecules, the hardware? Consider once again the human-made silicon computer. At the level of the electronic hardware, electrons act on electrons via physical forces following physical laws. This behavior is law-like. At the software level, we have symbols acting on symbols. Certainly the two levels of behavior are consistent. But can the law-like behavior of the electrons tell us anything about the purpose of the programs that run the symbols? No. Similarly, the causal laws of physics can tell us nothing about purposive behavior, the programs that biological forms, including the genes, obey, which have their source in consciousness and the morphogenetic fields.

Things get even more interesting, because this picture gives us something radically new: a solution to the age-old problem of defining a true biology of feeling.

Aside from all this fancy talk of law-like and program-like behavior, feeling is important because it's our everyday means to distinguish between life and nonlife. When you go to a restaurant and become curious whether the plant in the window is live or not, you touch it to get a "feel" that gives you the answer. By the same token, how do you know that you yourself are alive? You don't need a doctor's certificate to tell you that you are alive, do you? There is an inside feel to life that you are all too aware of.

Coming back to the morphogenetic fields, when we collapse their possibilities, the collapsed object of experience is feeling. What you feel is a sort of "energy" associated with the change or movement of those fields.

Quantum physics not only helps us find a nonlocal mediator, namely, consciousness, for the communication between the physical and the morphogenetic fields, but also helps expand biology to include feeling without phony ideas of adaptation. Both the physical and the morphogenetic fields are possibilities of consciousness. When consciousness collapses its possibilities, two parallel correlated experiences occur. One we call an experience of the physical world; this one we sense (or perceive). The other we call an experience of the world of morphogenetic fields; this one we feel. The two worlds do not interact directly, and dualistic issues don't arise. Instead the two worlds go on in parallel, and consciousness nonlocally maintains their parallelism (Goswami 1997b, 2004).

When Sheldrake introduced the concept of nonmaterial and nonlocal morphogenetic fields, he deliberately denied that he was reviving vitalism. I think the statement reflected a fear of outright rejection, because the word *vitalism* is anathema to many biologists. But I'll take a risk on the term, trusting that, in the decades since Sheldrake's radical proposal, frustration with the incompleteness of biology may have opened some minds. Let's replace the unwieldy appellation *nonphysical morphogenetic fields* with the much more friendly and familiar term "vital body," with the clear understanding that *the vital body is the container of morphogenetic fields* and therein lies its importance. And let's call the energies of the movements of the morphogenetic fields "vital energies." In the East these energies have been studied for millennia and are called by other names: *prana* in Sanskrit, *chi* in Chinese, and *ki* in Japanese. In the West, these energies have been called by various names over the centuries, but a recent term is *subtle energies.*

BRAIN, CONSCIOUSNESS, AND THE MIND

The psychologist C. G. Jung (1971) gave us an empirical classification of four personality traits—sensing, feeling, thinking, and intuition. Usually an individual's personality is dominated by one of these four kinds of experiences. What we sense belongs to the physical domain of consciousness. As argued above, our experience of feeling is due to nonmaterial entities—the morphogenetic fields of the vital body of consciousness. Next, let's consider thinking, which belongs to the realm of the mind. The questions are these: Is mind brain? And if mind is not brain, then what is the proper role of the brain in relation to the mind?

Biologists believe in the main that mind is an epiphenomenon of the brain arising from neuronal interactions. But recently, mind has been rediscovered in science as an independent entity; its purpose is to process meaning (Searle 1987, 1994; Penrose 1989). We now can generalize the previous discussion of the different types of consciousness possibilities and admit a new domain: a mental body of meaning. The proper role of the brain vis-à-vis the mind now suggests itself: The brain's role is to make representations of mental meaning (see chapter 15).

Returning to Jung's four traits, what about intuition? Well, before we can approach the idea of intuition, we must return to the idea of programs. Can the law-like behavior of the electronic hardware of a computer tell us anything about the laws by which the symbols operate on one another, the logic of the computer programs? Can the law-like behavior of the electronic hardware tell us anything about the programmer? The answers are again no.

These questions are a little too complicated to be dealt with in materialist metaphysics. We make an excellent start by adding the morphogenetic fields of the vital body: They are the blueprints of the vital functions that have to be represented in the physical form that is to carry out those functions. Biology is enhanced by including this organizing principle, but it needs several more. First, it needs the programmer. This element we have already introduced—consciousness and its agency of downward causation. Second, biology needs the nonphysical mind to explain how advanced biological organisms develop the capacity for meaning processing. Third, biology needs additional organizing principles that provide the context upon which the biological meaning programs operate, as well as the context for the biological functions that are mapped out by the morphogenetic fields.

The issue of context returns us to the question, Where do physical laws reside? This question has long occupied physicists and philosophers, and the most sensible answer is still the one given long ago by Plato. Physical laws are not written in the physical hardware, nor are they derivable from the random motion of the material substratum. Instead, as Plato says, physical laws are derived from the domain of archetypes, the most esoteric domain of the possibilities of consciousness. The laws of movement of the morphogenetic fields come from that domain. And it is also the source of intuition. Because this domain is beyond the mind, the term used is *supramental.*

Of course, materialists have quite a mental struggle with these archetypes. I am reminded of a *Dennis the Menace* cartoon in which Dennis is eating some cookies and telling his friend, "Mrs. Wilson says her cookies have love baked in 'em, but all I see are raisins." I should mention, though, that there is already some agreement about this subject among many scientists, biologists included, although the same scientists may have great difficulty in accepting a God capable of downward causation in "scientific" matters. The reason is subtle. These scientists have accepted a dichotomy of the world in order to reconcile their belief in both science and God. They do believe that God exists and is the ultimate creator of things in the sense that He lays down the laws that things obey (Miller 1999). But otherwise, they insist that God remains benign except maybe for occasionally helping out a devotee in a matter of need. This philosophy, called *deism*, allows these scientists both to go to church on Sunday and do materialist science the rest of the week.

In the same vein, the popular senator John McCain tried to make peace between creationists and evolutionists by saying that the two factions can find common ground by "letting the facts of evolution speak for themselves and letting the faithful see the hands of God in nature" (Winik 2005). Unfortunately, as our history shows, the dichotomous approach clearly does not work. We need a genuine integration. Fortunately, this is beginning to happen, as exemplified in this book.

PSYCHOPHYSICAL PARALLELISM

Jung's four types (1971), developed in his empirical study of human personality, describe the predominant mode through which a person experiences life. For some people the primary mode is the experience of physical sensation. Other people operate primarily through experiences of feeling, thinking, or intuition.

Each of these experiences—sensing, feeling, thinking, and intuiting—comes from a corresponding world of possibility— physical, vital, mental, and supramental (table 1). The possibilities collapse to actual events of conscious experience via conscious choice. There is no direct local interaction between these worlds in possibility, no dualism. Consciousness mediates their nonlocal interaction via simultaneous collapse. This is a new kind of psychophysical parallelism. In the old philosophy, the psyche and the physical world could exist in parallel, but dualism lurked there: What maintains the parallelism? By positing quantum consciousness as the mediator and maintainer of parallelism, this philosophy is made monistic.

Table 1. PSYCHOPHYSICAL PARALLELISM

Consciousness mediates the interaction between any two or more of the four bodies.

Human Experience (from Jung's typology)	Worlds of Possibility
sensing	physical
feeling	vital
thinking	mental
intuiting	supramental

Dualistic thinking comes naturally to us because of a difference in how we experience these four worlds. One world—the physical—we experience as external to us; the other three—vital, mental, and supramental—we experience as internal to us. Sometimes we refer to the physical world as "gross" because it can be experienced in a public fashion. We call the other three worlds "subtle" because we most often experience them as private. But neither conventional science nor any philosophy has ever explained the reason for this difference in our experiences of the various realms. Hence the dualism of internal-external, gross-subtle is created.

Quantum thinking allows us to find an explanation. The physical world has the microscale-macroscale division: Things at the microscale, which we can't perceive, make up things at the macroscale, the scale we can perceive through our senses. Quantum mathematics tells us that for a macro object, the quantum wave of possibility becomes very sluggish; it spreads in possibility, encompassing more and more possibilities with the passage of time, but at such a slow rate that between one person's collapsing it and another's, there would hardly be any change, and virtually the same actuality would come into manifestation. This sluggishness creates the illusion of identical sensory experience on the part of separate individuals who can share their experiences. This illusion gives rise to the notion that because the two people share an "identical" experience, the experience must pertain to a sensory object external to both people. Sensory, material objects are gross.

In contrast, the vital, mental, and supramental worlds have no micro-macro division. Objects in these worlds do not have a sluggishness arising from a composite nature, and thus they are capable of rapid change; that is, the quantum possibility wave spreads rapidly. Between one person's experience of objects in any of these three worlds and another person's, the quantum possibility wave of the object would change so much that, generally speaking, two people would collapse different experiences. Because their experiences of these worlds are thus private, they conclude that these worlds are internal to each of them, that these worlds are subtle.

The gross physical world, with its micro-macro and quantum-Newtonian (or approximately Newtonian) division, is needed as a ground upon which to make representations of the subtle worlds. Even more importantly, the gross physical world is needed to make manifest experience (see chapter 5).

SITUATIONAL CREATIVITY AND FUNDAMENTAL CREATIVITY

Let's return to the subject of the supramental, this time in connection with creativity. Creativity, as we ordinarily experience it, is mental creativity, a novel invention or discovery in meaning. Subsequently, we make a physical representation (a product) of the new meaning. But a subtlety is present that now can be explicated.

One of the most amazing things about creativity is that it carries a built-in feeling of surprise, so a creative experience is sometimes called an "Aha!" experience. This surprise has to do with a fundamental discontinuity in the creative process, a discontinuity that elsewhere I have identified as a quantum leap in meaning (Goswami 1999).

Mental meanings can concern not only the physical and the vital worlds, but also the (mental) representations of the supramental archetypes. As the contextual environment (physical, vital, mental) of thoughts changes, problems arise, and discovery of new meaning occurs in the form of solutions to the problems. If the new meaning that answers a problem is found by reshuffling old meanings in existing supramental contexts, we call the creative insight an invention and the process one of "situational" creativity. For situational creativity, one does not need to step out of the arena of the mind.

Sometimes, however, to find the new meaning that answers the problem—or just formulates it properly—we need to find a new supramental context. That is, we need to take a fresh look at the supramental domain of archetypes. This examination of course requires a quantum leap from the mental to the supramental; the resulting creative insight is called discovery, and the process is one of "fundamental" creativity.

To repeat, situational creativity is the invention of new meaning within an old archetypal context or within a combination of old archetypal contexts. Fundamental creativity, on the other hand, is the discovery of not only new meaning but also of a new archetypal context, accomplished through a quantum leap from the mental to the supramental.

To see the distinction clearly, consider two of Darwin's insights. Darwin began his life as a believer in Christianity and creationism. For him to realize that environmental adaptation by organisms through geological time must be the result of evolution required a discontinuous quantum leap, a change in archetypal context of his thinking from a religious worldview to a modern worldview. This is an example of fundamental creativity. On the other hand, when Darwin read Thomas Malthus' work on the danger of overpopulation and the importance of competition in our struggle for survival and formulated his idea of the survival of the fittest in biological evolution, he was working with new meaning within known contexts; his creativity was situational creativity.

Biological creativity is creativity in the vital domain, rather than the mental domain, but the same considerations apply (see chapter 8). As the physical environment changes, the existing vital blueprints and their physical representations are no longer adequate to carry on necessary biological functions. This crisis represents a certain threat to survival, but it also represents an opportunity for consciousness to improve its repertoire of vital blueprints and their physical representations. Thus does biological creativity come into play.

THE BIOLOGICAL
ARROW OF TIME REVISITED

Let's revisit the biological arrow of time, the observation that time seems to flow from the past to the present, into the future. Biologists who try to base their science firmly on physics also have to face another "hard" problem. Time in physics has no direction, or rather both directions, backward and forward, are equivalent. Fundamental physical laws of movement remain the same if you change the direction of time from forward to backward. As discussed in chapter 1, physics explains the arrow of time by the phenomenological concept of entropy and the entropy law— entropy always increases. By empirically looking at the entropy of the universe, we can distinguish the past from the future. The physical arrow of time is an arrow of entropy.

Biology also gives us an empirical arrow of time. Biological organisms evolve toward increasing complexity. But biological complexity consists of more order and thus is the opposite of entropy, which is disorder. So the two arrows are not compatible. Furthermore, Darwinism fails to give any explanation of how the biological arrow of time arises. According to Darwinism, evolution can go either way, to simplicity or complexity.

With the ideas developed thus far, this problem becomes tractable. Biologists have been missing something about evolution because, misguided by Darwinism, they have been reluctant to admit the role of purpose in evolution except in the dubious context of adaptive behavior. When we study the purposiveness of evolution from the framework of biology within consciousness, we can readily see the explanation of the biological arrow of time. This purposiveness is the subject of the next chapter.

5

*E*VOLUTION:
From *W*HAT *to* *W*HAT?

*T*o the spiritual traditions, Jung's four ways of being, or as I am calling them here the four domains of con-sciousness, look like the progression shown in figure 4:

SUPRAMENTAL
(the domain of godly archetypes)

↑

MIND
(the domain of meaning)

↑

LIFE
(the domain of vital morphogenetic fields)

↑

MATTER
(the domain of manifest representation making)

Figure 4. The four domains of consciousness as
viewed by the spiritual traditions.

This hierarchy resembles what the philosopher Ken Wilber calls the great chain of being. Biologists denigrate the great chain of being. They see life, mind, and godliness, all three, as adaptive epiphenomena of matter with no consequences other than survival benefit. I hope the discussion in the previous chapters has helped you to arrive at some healthy skepticism about the materialist premise that life, mind, and consciousness can evolve in matter without the help of other organizing principles. But then the puzzle of the progression above remains.

If science is based on the primacy of consciousness, then the important question is this: If consciousness is the ground of all being, "point alpha" for evolution, then how does evolution proceed? The answer is enlightening. It was first given by philosopher Sri Aurobindo in 1939 (Aurobindo 1996) and the Jesuit priest Pierre Teilhard de Chardin (1961). Later Ken Wilber (1981) and I (Goswami 2001) both elaborated upon the idea.

We note that in the beginning consciousness includes all possibilities. Ponder what that means. Among other things, "all possibilities" must include literally *all* possibilities, past, present, and future. In other words, when every possibility is included, there is no scope for the passage of time. To bring time into the equation of the manifest universe, consciousness must limit what is possible. The imposition of progressive limitation on what is possible is seen as an involution of consciousness. In this way, when evolution is viewed from the context of the primacy of consciousness, involution must precede it.

From a primacy-of-consciousness point of view it is also possible to ask, What is the purpose of evolution? Why evolution at all? The answer is easy: Evolution is needed for experiencing the possibilities of consciousness in manifestation. When consciousness is inseparable from its possibilities, there is only one thing, and no experience is possible. As the mathematician G. Spencer Brown (1977) pointed out, "We cannot escape the fact that the world we know is constructed in order (and in such way as to be able) to see itself, but in order to do so, evidently it must first cut itself up into at least one state which sees, and at least one other state that is seen."

Awareness, that is, a subject-object split, is needed for experience. With this point agreed, the purpose of evolution can be stated, following a statement by Carl Jung. The purpose of evolution is to make the unconscious (that of which we are unaware) conscious (that of which we are aware and capable of manifestly experiencing).

Once we recognize that the enfolding and the subsequent unfolding of the possibilities of consciousness are a purposive play, we can easily lay out the stages of the involution, the imposition of limitation, that precedes evolution. When we start a game, how do we start? We lay out the rules.

Some people find it a little hard to comprehend why an "omnipotent" God would play the game of creation within rules. But play without rules is no fun for us, and we are made in God's image. So why should God's wanting to play the game within rules be so surprising?

I am reminded of the comedian Bill Cosby's Noah routine. God speaks to Noah from up above; he is giving Noah instructions for (of course) building the ark. But Cosby's modern Noah is a little skeptical: "Who is this, really?" The voice insists, "This is God." But Noah still cannot believe: "Am I on *Candid Camera?*" He starts looking for the hidden camera. Now God appears in a convincing form and starts again with serious instructions for building the ark. But Noah raises that issue of omnipotence. "If you want this ark so much, why don't you build it yourself?" And Cosby's God says what any scientist would approve. "Noah, you know I don't work that way."

God plays within rules; otherwise science would be impossible. (But of course these rules can be and are subtle, like the quantum rules that permit a certain amount of creativity.)

So the first stage of involution is the imposition of a set of rules; call it a body of laws if you wish. It contains the laws of behavior of all that will be the case subsequently, including the laws for its own operation. We will recognize this body of laws as the supramental body of consciousness introduced in the last chapter.

I hope you are getting the idea. The next limitation is that the game is played with only those possibilities that are meaningful, in this way creating the meaning-processing mind.

The vital energy body imposes further limitation on the possibilities that come into play: Only those possibilities pertaining to certain body plans or blueprints or morphogenetic fields of biological being, that is, life, are allowed. The final limitation is the requirement of manifestation, the collapse of the possibility waves belonging to all the domains, including the final one that we call "matter," which is needed for making representations of what went before.

THE GROSS (MATERIAL) BODY
AND THE SUBTLE BODIES

As Descartes noted early on, mind (and by implication all our other internally experienced bodies, the vital and the supramental) is without extension, but matter has extension. This means mind is just one thing, whereas matter is reductive, that is, the micro makes up the macro. This state of affairs is necessary so as to make manifestation possible. Let's see how.

In chapter 2, I noted that consciousness becomes self-referent in us in the act of quantum observation, in the act of choosing actuality from the quantum possibilities. Let's discuss this aspect of consciousness in some detail.

You may not have noticed, but we can see paradox in the observer effect in another way. The observer chooses, out of the quantum possibilities presented by the object, the actual event of experience. But before the collapse of the possibilities, the observer himself (or herself) consists of possibilities and is not manifest.

So we can posit the paradox as a circularity: An observer is needed for collapsing the quantum possibility wave of an object; but collapse is needed for manifesting the observer. More succinctly, no collapse without an observer; but no observer without a collapse.

If we stay in the material level, the paradox is unsolvable. The consciousness solution works only because we posit that consciousness collapses the possibility waves of both the observer (that is, his or her brain) and the object simultaneously from the transcendent reality of the ground of being that consciousness represents.

The artificial intelligence researcher Douglas Hofstadter (1980) gave us the clue for understanding what is going on. Such circularities as that above, he noted, should be called tangled hierarchies, in which the causal levels are infinitely intertwined. Most interestingly, self-reference, the subject-object split, emerges from such circularities.

Let's consider an example given by Hofstadter. Consider the liar's paradox, expressed by the sentence "I am a liar." Notice the circularity: If I am a liar, then I am telling the truth; if I am telling the truth, then I am lying, ad infinitum. However, this infinite oscillation has made the sentence very special: The sentence is speaking of itself, separate from the rest of the world of discourse.

But this apparent separation of the self of the sentence and its world, its apparent self-reference, depends on our understanding and staying within the rules of English grammar. The circularity of the sentence disappears for a child, who will ask the speaker of the sentence, "Why are you a liar?" Grammar, although the real cause of the paradox, is implicit, transcending the sentence.

In the observer effect, this same issue explains why it took us physicists a while to decipher the situation. The choosing quantum consciousness—God, if you will—is implicit, not explicit; transcendent, not immanent. The collapse is tangled hierarchical in nature, giving the appearance of self-reference and a subject-object split. The observer-I, the apparent subject of the collapse, arises codependently with the object because consciousness becomes identified with the observer's brain but not with the object (Goswami 1989, 1993).

There is something special about the brain. To see this, consider a Geiger counter. This device is usually thought of as a measurement apparatus, but in truth it only amplifies the incoming stimulus or signal. A Geiger counter is simple hierarchical: What is signal (the micro) and what is the amplifier being affected by the signal (the macro) are clearly distinguished; thus they form a simple hierarchy of cause moving one way. By contrast, in a self-referential system such as the brain, the causal levels involved in the transition from micro to macro form a tangled hierarchy (see chapter 15 for further details).

Whenever a quantum possibility wave is collapsed, a tangled hierarchy is present in its measurement. But the existence of the tangled hierarchy depends crucially on the micro-making-up-macro aspect of matter. Thus quantum collapse and thus manifestation can take place only when matter comes into play.

It is worth noting that the notion of tangled hierarchies can shed light on a perennial issue in the debate over intelligent design. Critics of the intelligent designer idea seem always to be puzzled by the question of who designed the designer. They think it is impossible to answer this question, although the mystics of the world noted long ago that the designer must be capable of self-creation, by which they meant self-reference. In Buddhism this point is noted very clearly in the doctrine of dependent coarising of subject and object. Now, after many millennia, we are beginning to understand how the self-reference comes about and giving a reasoned foundation to the mystic's intuition.

Returning to our central concern, we have seen that evolution begins with matter, or, more precisely, with the first living cell carrying out tangled hierarchical quantum measurement (see chapters 7 and 8 for a discussion of this statement). We next must ask, What is the future of evolution? How does evolution end?

The Stages of Evolution

The stages of evolution can now be codified. But the progression of evolution as described in the new science is quite different from that in materialist biology. What evolves? In the present view, matter doesn't evolve into life, but rather the whole material universe evolves in possibility until the first living cell and its environment are ready to manifest the rudimentary biological functions (reproduction and maintenance, including waste elimination) and to carry out tangled hierarchical quantum measurement. The evolutionary journey of life now begins: the evolution of the capacity of matter to represent the vital morphogenetic fields.

The evolution from simplicity to complexity that we observe in the evolution of life can now be explained. First, complexity permits increasingly sophisticated representation of the morphogenetic fields that have already been evoked. Second, complexity makes room for forming representations of previously unrepresented morphogenetic fields that correspond to "higher" biological functions. Clearly the second outcome also creates more complexity.

With an understanding of the evolution toward complexity, the biological arrow of time is no longer a mystery. As organisms get more sophisticated as a result of evolution, they represent within themselves more and more morphogenetic fields, and the fields are represented with more and more sophistication. Over the course of this change, the organisms become more sophisticated in processing feeling. And all this creation of complexity, this increasing order and sophistication, requires the involvement of creativity from consciousness.

While life's arrow of time consists of developing more order, more complexity, how does life's environment evolve? When there is no creative demand, consciousness chooses according to the probabilistic laws of quantum physics. This mode of choice produces increasing randomness and entropy. Therefore the environment's entropy continues to increase. How is life sustained, then, against this prevailing current of increasing entropy? Life can only flourish in the presence of a large source of negative entropy (Schrödinger 1944). For us, this large source of negative entropy is the Sun.

Eventually, when the brain evolves, the organism develops the capacity of representing the meaning-processing mind. This step begins a new era of the biological arrow of time—the evolution of the capacity for making mental representations. This aspect of evolution has gone unnoticed by biologists, but fortunately it has been quite extensively codified in anthropology and sociology. Over time, meaning processing has become more and more sophisticated and more and more universal. Right now, we are in the middle of this mental era of evolution.

By noting the nature of this second phase of the biological arrow of time, we now can easily project what the future of evolution must hold for us.

THE FUTURE OF EVOLUTION

In neo-Darwinism, talking about the future of evolution is fruitless. The following lines from "Evolutionary Hymn" by the romantic children's fantasy writer C. S. Lewis reflect this sentiment just about perfectly:

> Lead us, Evolution, lead us,
> Up the future's endless stair:
> Chop us, change us, prod us, weed us.
> For stagnation is despair:
> Groping, guessing, yet progressing,
> Lead us nobody knows where. (Lewis 2002, 55)

The reason nobody knows where is that in the Darwinian two-step mechanisms of chance and necessity, there is no progressivity, no tendency toward increasing complexity; all is directed toward survivability. The latter depends primarily on fecundity, or how much progeny an organism leaves behind, and not much else.

But in our model of evolution preceded by involution, one more stage of the evolutionary arrow of time can be easily predicted. There must be a third phase consisting of the evolution of the capacity of matter to make representations of the supramental archetypes. Let us see what that means.

TOWARD THE OMEGA POINT

Can we ever represent the archetypes of the supramental directly in the physical without the help of the intermediaries, the blueprints, the mind and the vital body?

To answer that, I would like to begin by telling you about the work of the visionary Jesuit monk and biologist Teilhard de Chardin (1961). Everybody knows about the biosphere—the web of organic life around and about the material planet Earth. Teilhard de Chardin saw the second phase of life's evolutionary arrow of time as a noosphere, a sphere of the evolving mind of humanity. The noosphere exists in the outer dimension of the world, in our civilization, in our books and in the Internet, but not solely there. The noosphere also has an internal dimension.

Teilhard de Chardin envisioned that when this inner dimension of the noosphere, right now quite fragmented, somehow fuses, thereby taking on an internal dynamics of its own, then there will be a new phase in our evolution, a quantum leap that he called the omega point.

Sri Aurobindo (1996) had the same vision. He saw the human mind going through a number of evolutionary phases, still all brain based. Then, with the development of what he called the overmind, the brain-based humans reach a pinnacle. At the next stage, superhuman beings evolve who are able to manifest the godly qualities, the supramental archetypes of the mind, directly in matter. "Just as the animals are the laboratory for the human, the human is the laboratory for the superhuman," he said.

In the schema that I have represented in figure 5, the vision of both Teilhard de Chardin and Aurobindo will come to fruition when evolution produces tangled hierarchical vehicles of matter capable of representing the objects of the supramental domain—the archetypes of the vital and the mental bodies—directly. This is the glorious future of our evolution.

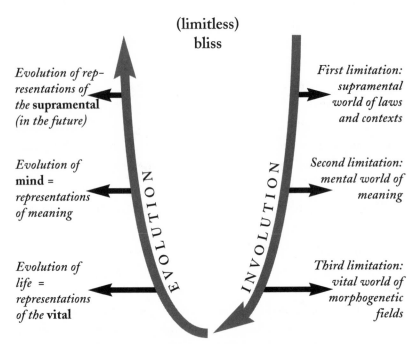

Figure 5. Involution and evolution.

Think about what it means. Today, some of us take quantum leaps of creativity to glimpse the supramental. Then we make vital and mental representations of the insights gained from those leaps. And then if we are so motivated we live these vital and mental representations to make circuits in our brain, so these modes of godly living become habits of transformed behavior. We call such transformed people saints and sages.

But these brain representations are all secondary: They are made from vital and mental representations. It is like drawing a picture from someone else's description. In this way, one sage's living representations of godliness differ from another's enough to cause confusion among the followers, especially when the teachings of these sages are translated by lesser people who haven't bothered to transform. Thus we create religions and differences in our views of God and godliness. These religions fight among themselves and also with science. This has been our big problem throughout human history, hasn't it?

But suppose a direct physical representation of the archetypes were available for everyone to check as needed. Such a representation would remove all need to fight about whose mental representation is most valid. Also, if we had the benefit of this privileged view, then we could teach supramental qualities, just as today we can teach our students mental concepts of our insights without their having to take quantum leaps and manifest the insights personally. In this way, we evolve rapidly toward unity within the diversity that began at the alpha point.

The Rest of This Book

In these first five chapters, I have given you a racing overview of the scope of the new biology, which can become a complete biology once we give up the prejudice for materialism and reductionism and, instead, base biology on quantum physics and the primacy of consciousness. You may feel you've overdosed on too many new concepts. But I promise that in the rest of the book I will take you slowly through these concepts, and then they won't appear as difficult or strange. I hope they will, in fact, appear very friendly and satisfying, because you will see that with their help all the problems and controversies of biology can be quietly laid to rest. Here is a tour of the remainder of the book.

In part 2 (chapters 6–8), I will discuss the solution of the hitherto unsolved mystery of the origin of life. Part 3 (chapters 9–12) delves into the central theme of the book: creative evolution and its motivation, history, dynamics, and relation to Darwinism. In part 4 (chapters 13–14), I elaborate further on the vital body of the morphogenetic fields and explore the new avenue it opens in biology—feeling. Chapter 14 also explores independent evidence for vital energy, including alternative medicine. In part 5 (chapters 15–17), the role of the brain in biology-within-consciousness is investigated. The so-called mind-brain problem is solved, and the neurophysiology of perception is explained in a paradox-free manner. Chapter 17 discusses the evolution of the representation making of the mind through the current era into the future. In part 6 (chapters 18–21), the main subject is the supramental contexts of evolution. First, in chapter 18, we explore the subject of death and survival after death; in chapter 19, we explore the questions of deep ecology and bioethics. A timely replacement is found for the current sociobiological bioethics based on the "selfish gene." Chapter 20 celebrates a revival of Lamarckism with an explanation of the hitherto unexplained phenomenon of instincts. The book ends with a discussion of our immediate evolutionary future and what we can do to facilitate its course.

PART 2

LIFE *and*

ITS ORIGINS

6

MATERIALIST THEORIES of LIFE and ITS ORIGIN

The question, What is life? reminds me of a Buddhist story. An inquisitive fellow creates havoc among the gods in heaven by asking, "Where does the universe end?" Finally, the creator himself, God, appears. When the fellow repeats his question, God takes him aside and says, "These gods look to me with such reverence that I could not say this in their presence. You see, even I do not know where the universe ends."

In my mind's eye, I visualize a modern fellow going to the biology department of a famous university and creating havoc by asking, "What is life?" Some of the professors answer that life is self-maintenance, which is a function of proteins; some say it is self-reproduction, for which DNA is the main player. But the fellow is not satisfied: He counters, "A crystal can maintain its form, and a fire on the prairie tends to reproduce itself, but we don't call those systems alive, do we?" Neo-Darwinists come along and say, "Life evolves; evolution is what separates life from nonlife." But the fellow is patient, and he counters again, "The universe also evolves. Would you call the universe alive?" Some more radical people, conversant with physics and chemistry, chime in, saying that life is a self-organizing system. But the fellow won't shut up; he is quite radical himself, you see. "Certain chemical reactions consisting of autocatalysis have a self-organizing property, but nobody would claim that these chemical reactions create life from the nonliving," he insists. Finally, the head of the department herself arrives, takes the fellow aside, and says, "My colleagues hold me in such esteem that I did not want to admit this in public. Also any public admission of the shortcomings of biology would only put more ammunition in the hands of those creationists. More court cases, who needs them? But you see, we biologists still do not know how to define life, not by any one defining characteristic, anyway. So we manage by listing a whole bunch of properties as characteristics of life, hoping that no nonliving system will ever be found with too many common characteristics with the living to create confusion."

What are these characteristics? In one biology textbook (Minkoff and Baker 2004), I noticed the following plethora of properties mentioned in defining life. I paraphrase them here:

- **Organization:** Living systems are organized starting with the cells, to multiple cells or tissues, to organs, to the organism.

- **Metabolism:** Living systems are capable of maintenance through metabolism—taking in energy-rich material and eliminating, on the average, lower-energy material. Some of the energy acts as the fuel of living processes; some accumulates and is released at death.

- **Homeostasis:** Living systems can maintain homeostasis even under threatening environmental conditions by at least partly converting toxic chemicals into less harmful ones.

- **Selective response:** Living systems can respond to stimuli selectively. They can identify what is food and can move away from offensive stimuli.

- **Growth and biosynthesis:** Living systems go through phases in which they take in material from the environment and grow.

- **Genetic material:** Living systems contain hereditary information in the form of genes (portions of nucleic acid molecules—DNA and RNA) from previously living systems.

- **Reproduction:** Living systems can reproduce new living systems similar to themselves by transmitting some or all of their genetic material.

- **Population structure:** Living systems (organisms) form populations—groups related by common ancestry. Organisms are capable of sexual reproduction, and the organisms of a population can interbreed.

You can see that this is altogether a clumsy definition, but for the true materialist, this is as good as it gets. Let's first review some models developed by both biologists and nonbiologists for the origin of life and appreciate the difficulties these models face.

REDUCTIONIST MODELS FOR THE ORIGIN OF LIFE: SO CLOSE AND YET SO FAR

Many biologists today have convinced themselves that life is defined as the molecular biology of the living cell and no more—or possibly even less. I want to give you a feel for how difficult it is to make a plausible materialist model of the origin of a functioning living cell.

To appreciate these models, you have to appreciate some of the complexities of the living cell. Single-celled creatures are of two kinds: prokaryote (without a nucleus) and eukaryote (with a nucleus). I have already mentioned the two most important components of a cell: DNA and proteins. A third important component is the ribonucleic acid (RNA) molecules, which mediate information exchange between DNA and proteins. A fourth component is the cell membrane, which gives the cell a defined structure: an inside and an outside. The cell membrane is made of fatty acid molecules, such as lipids. A fifth component is the cytoplasm, which fills the cell body, facilitates transport of proteins, and acts as a solvent (fig. 6a). In addition to all of the preceding, eukaryotic cells also have a nucleus. That is, the DNA and the machinery that conveys DNA information to RNA are sequestered by a distinct membrane: a cell within a cell, so to speak.

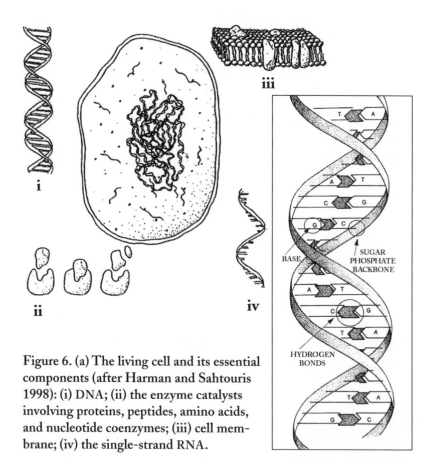

Figure 6. (a) The living cell and its essential components (after Harman and Sahtouris 1998): (i) DNA; (ii) the enzyme catalysts involving proteins, peptides, amino acids, and nucleotide coenzymes; (iii) cell membrane; (iv) the single-strand RNA.

(b) The DNA double helix in more detail.

Let's discuss the complexities of some of these cellular molecules. A protein is a large polymer chain of amino acid molecules, so complex that no one can conceive of ever succeeding in manufacturing it in the laboratory starting from the amino acids. In the cell, of course, DNA has the code (the genetic code) that directs the assembly of the amino acid sequences to make proteins.

DNA is famous for its double-helix structure (fig. 6b). The two strands of the double helix are joined chemically through the electrical bonding of the nucleotide molecules adenine (A), thymine (T), guanine (G), and cytosine (C). The genetic code is contained in the sequences of the letters along one strand, divided in groups of three letters.

Just as DNA has the code for the production of proteins, the proteins are instrumental for producing DNA. This circularity—in which DNA makes proteins (with the aid of RNA) and proteins make DNA—prompts origin-of-life researchers to reject DNA as a candidate for primitive life. It's just too complex.

The most popular models assume that life began with the self-replicating, single-strand nucleic acid chain, the RNA. Could an RNA molecule come about by chance in a primordial soup of the basic elements of life?

In the fifties, Stanley Miller and Harold Urey conducted a famous experiment (for an account, see Miller 1974) in which amino acid was synthesized from a "primordial soup" of the basic elements carbon, nitrogen, oxygen, and hydrogen by passing energy through the soup.

However, no experiment of the Miller-Urey type was immediately successful for the synthesis of RNA. But in the sixties, the biochemist Sol Spiegelman made history with a successful experiment suggesting that if a protein enzyme were present, the catalytic effect of the enzyme would be of enormous help in synthesizing RNA. (Catalysis occurs when a chemical substance, the catalyst, enhances the rate of a chemical reaction but is itself regenerated at the end of the reaction.) "Enzymes can do anything" is the popular wisdom in biochemistry.

Unfortunately, to manufacture a protein enzyme by blind chance is not easy either. An argument based on probability calculation shows that the manufacture of even a relatively small protein enzyme would take a time much longer than the age of the universe.

To see why, we will make a little digression for the mathematics-minded skeptic. Take a look at figure 6b, the DNA molecule. It takes three of the base molecules (denoted by the letters A, C, G, and T) in a given sequence (called a *codon*) to make an amino acid molecule. Consider, following the astrophysicist Arne Wyller (1999), the probability of a protein enzyme consisting of a chain of 34 amino acids being put together by chance. This sequence would have 3 times 34 or 102 base molecules or letters.

To get the first letter of the sequence right, the probability is ¼, because there are four letters to choose from. To get the second one right, the probability is again ¼. And to get the two of them right together, the probability is ¼ times ¼ or $\frac{1}{16}$ (probabilities multiply).

You get the picture. To get 102 letters right, we need to multiply ¼ by itself 102 times, or $\frac{1}{4}^{102}$. If we want to do this protein right by trial and error, chance, we have to try 4^{102}—or 10^{61}—times!

How long does each trial take? Wyller makes the point that even if we take a minuscule 1 second for each trial, even then the time taken is 10^{61} seconds. Given that the life of the universe is about 10^{18} seconds, the time to make a simple enzyme by trial and error exceeds by many times the entire life of the universe. The time available for the primordial soup to make life is roughly equal to the time between the appearance of the Earth's oceans and the first life form, a measly 400 million (4×10^8) years.

This mathematical digression has solidified our conclusion: We cannot expect enzymes to help the manufacture of the primitive life forms of RNA. Instead the proponents of RNA play with an idea from game theory known as the prisoner's problem.

A prisoner is planning to escape from his prison cell (fig. 7) with the help of a friend from outside. Obviously, escape is easier if they both dig from opposite sides of the same corner. Unfortunately, no communication is possible, and they have six corners to choose from. The chance of escape doesn't look so good, does it? But if you put yourself in the prisoner's position, the chance is excellent that you will dig from corner 3. Why? It is the only corner that is convex from inside. Also, chances are excellent that your friend will notice the same thing and decide to dig there also.

Now ponder a moment. What is your friend's motivation to actually dig at corner 3? Your friend's motivation is you, the expectation that you will dig there.

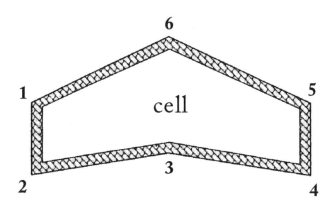

Figure 7. The prisoner's problem: which corner to dig at?

This is called a feedback system. Notice also that this kind of feedback increases the tendency to move away from the state of equilibrium in which nobody does anything and the status quo is maintained. This is a positive feedback system.

A chemical reaction, called the *Belousov–Zhabotinsky Reaction* after its discoverers, displays a positive feedback loop. The details aren't necessary, but here's the gist of why this reaction is different from ordinary chemical reactions. In an ordinary reaction, you put some chemical ingredients in a dish; they mix and react and the product sits peacefully at the bottom of the dish, like a marble at the bottom of a spherical bowl. But in a Belousov-Zhabotinsky reaction, depending somewhat unpredictably on the initial conditions, the system may set up as an oscillating chemical clock, or it may develop into waves of chemical disturbance spreading over a macroscopic distance, or it may just display a beautiful spatial pattern.

The mystery at the heart of the Belousov-Zhabotinsky reaction is cross catalysis. Cross catalysis is the process in which two or more chemicals (the cross catalysts) mutually produce one another in a cyclical fashion while enhancing the rate of certain reactions. What is interesting here is the ability of the system to continuously recreate itself, a phenomenon called *autocatalysis*.

Biochemists play with the idea that the RNA, the first life, came about through some sort of autocatalysis. In this theory, proteins also got into this game somehow (never mind the low probability for this happening). Opinion varies as to whether initial proteins helped the formation of the replicator RNA or whether the RNA helped to manufacture proteins. At the next stage, the genetic code developed in some primitive form, along with some primitive form of DNA, and the nucleic acids and proteins participate in many primitive "hypercycles" (inclusive cycles). In the final stage, fatty acid (lipid) molecules made an enclosure (the cell wall) to make the complete, living prokaryotic cell.

Biologists such as Walter Gilbert and Manfred Eigen (see Eigen and Schuster 1979) are the famous players of such models. But the nagging problem of low probability persists at each stage of their scenario for life's synthesis, so biologists further play with the controversial idea that the Darwinian mechanism of chance and necessity can be invoked to boost the probabilities. This step is controversial because Darwin's theory is supposed to apply for the evolution of life after it originated. Can it be invoked in pre-life scenarios? Moreover, small probability is only part of the problem. Within the model itself, conflicting mechanisms narrow the range of applicability of the model, forcing consideration of other scenarios (for a review, see Casti 1989 and Davies 1999).

One such scenario posits that proteins came first, then DNA. I stated above that DNA (acting through the mediation of RNA) specifies the order of amino acids that are used to construct the protein molecules. Information flows one way, from DNA to protein. This statement captures the central dogma of molecular biology. But the biologist Tom Cech and collaborators experimentally demonstrated that an enzyme called *reverse transcriptase* can catalyze the synthesis of DNA from an RNA template. Therefore, why not theorize that proteins came first in the game of the origin of life? In fact, experiments of the Miller-Urey type do show that protein-like chains of amino acids (we can think of them as primordial proteins) can be formed easily by passing an electrical spark through a closed vessel containing methane, ammonia, hydrogen, and water, a mixture simulating the primordial environment of the Earth. Theorists working in this vein propose that proteins came first, forming spontaneously. After a while, something like reverse transcriptase created DNA. But these models, too, face many criticisms, low probability being one.

Another scenario is that life began twice: once as protein, and then again as protein plus nucleic acids, as in the present-day cell. In still another scenario, instead of assuming a hot, watery soup for the primordial environment, the honor goes to hot clay, and a crystalline life is seen as the precursor of organic life as we know it. This model became popular with nonscientists because of its biblical flavor but did not gain much favor among origin-of-life researchers.

It is said that after his momentous (co)discovery of the double-helix structure of DNA, Francis Crick went to a pub (to celebrate?), where he promptly declared, "I have the secret of life."

Perhaps all the origin-of-life theorists take Crick too seriously in proposing, over and over, that life must have begun as either DNA or protein. But in truth, can we believe that a bag of DNA or proteins would behave as a being that is alive in any real sense? I think not.

As you can see, much of this is play: Making plausible scenarios for life's origin can be fun. But infusing your playful scenario with enough plausibility for others to take it seriously is a difficult task, and no materialist model has succeeded yet.

Reductionists know that if you cannot solve a problem as is, then it's wiser to redefine it and solve the redefined problem. The researchers I have mentioned so far were trying to make plausible models for the origin of life as we know it—the DNA/RNA protein conglomerate confined within a cell wall with myriad other molecules to support cellular functioning. Suppose we simplify this down to the molecular level and demonstrate that Darwin's chance-and-necessity-driven evolution begins to work at that level. Could we not then claim that our problem is solved? For those who believe that Darwinian chance and necessity can evolve consciousness in nonsentient beings, it certainly seems plausible that those forces can evolve a living cell out of raw molecules.

Three experiments, two performed decades ago and the latest performed in the nineties, seem to support this view. Let me begin with the experiment by Spiegelman already mentioned (Spiegelman 1967). Spiegelman's experiment was based on a small RNA virus named Q_{beta}. Viruses, which are a strand of DNA or RNA encased in a coat of protein, store genetic information but cannot replicate on their own. They have to invade and occupy cells, hijack their reproductive apparatuses, and adapt them to replicate more viruses. Spiegelman demonstrated that Q_{beta} does not need a living cell but can replicate in a test tube if it is allowed to incubate in a medium containing its own replication enzyme, a supply of raw materials, and some salts. Spiegelman then put some of the freshly synthesized naked RNA in another test tube of nutrient solution, and let it multiply. Then he used some of the new RNA of this batch in another nutrient solution to manufacture yet another batch of RNA, and so on in a series of steps.

Spiegelman's naked RNA strands did not manufacture protein; they just replicated. The effect of allowing unrestricted replication was that the RNA that replicated fastest got passed on most effectively to the next generation, thus simulating a Darwinian competition. In 74 steps, the RNA strand went from 4500 nucleotide bases to only 220 bases. This one, the survivor, was called Spiegelman's monster.

The second experiment was carried out by Manfred Eigen and his collaborators in 1974 (see Eigen et al. 1981), who also started with a chemical solution containing Q_{beta}, its replication enzyme, some salts, and an energized form of the four bases that make up the RNA nucleotides. However, these experimenters began reducing the quantity of viral RNA added to the mixture at each step. As the amount of input RNA was progressively reduced, with reduced competition the RNA enjoyed vigorous exponential growth, and the exponential growth continued down to the input of a single strand of RNA. And then the surprise! Even without the input of a single initial RNA, replicating RNA strands were produced. In other words, RNA strands—self-replicating molecules—were produced in the laboratory for the first time.

If you believe that viruses are alive and self-replication is the definition of life, this experiment almost shows the creation of life. However, most biologists would not go so far in their claim. They would recognize that the replicator enzyme is still borrowed from living material. Eigen's experiment is a very intelligent piece of bioengineering, the manipulation of biological information by means of human intelligence.

In the 1990s, a third set of experiments was performed. The goal was to produce small, simple replicator molecules in cell-free media and then subject them to environmental stresses intended to induce mutation and evolution. The conclusions from these studies are startling, as summarized by a researcher on the subject at a conference and paraphrased here:

- It was found that polynucleotides (polymer chains of nucleotides) including RNA can replicate in the presence of suitable enzymes in a test tube, without the confinement of cell walls.

- Related small molecules called *oligonucleotides* were found to be replicable even without the assistance of proteins or enzymes.

- Molecular replicators have been found outside the nucleotide family as well.

- Molecular replicability can then be taken as a physical measure of fitness in the Darwinian sense, opening the door for evolutionary processes to be described and analyzed in the language of physics.

- In particular, Darwinism's tautology problem seems to be solved. This problem refers to the embarrassing situation that Darwinism presents to us: Who survives? The fittest. Who is the fittest? One who survives. But now the definition of fitness in terms of molecular replicability can be given independently from survival; hence, no tautology is involved.

- The usual argument against a Darwinian mechanism of evolution, namely, small probability, is claimed to have been negated for evolution experiments with the RNA molecules. The negation is based on the experimental proof that target-oriented adaptation indeed is feasible.

But a closer examination of the results may quiet some of the hoopla. That humans have produced a situation in which molecules can replicate, mutate, and evolve in the test tube simply means that human intelligence has managed to establish a physicochemical basis for self-replication. This condition is very far from the condition that needs to be proven—that inanimate, primordial nature can do the same thing through natural processes without help from intelligent beings, outside causal agents. It will not do to say that we are part of nature, too. Furthermore, self-replication of a molecule is not the same thing as self-reproduction of a living cell, a conglomerate of many molecules.

On top of this, the big question remains. It is very premature to assume that once Darwinian chance and necessity come into play the rest of the production of life is a cinch. It is the contention of this book (and a position supported by many biologists) that the Darwinian mechanism is quite inadequate to explain the short-tempo events (the punctuation marks) of macroevolution. If this is so, how credible is it to assume that the Darwinian mechanism is adequate to take us from simple replicable molecules to the full-fledged, self-reproducing, living cell with its genetic code and DNA-protein hypercycle?

Furthermore, we should not be too hasty in thinking that solving Darwin's tautology is a big breakthrough, because this so-called breakthrough also exposes a big weakness of Darwin's theory. Darwin's law of evolution through chance and necessity may not be a genuine biological law, after all. Instead, it may be a corollary of a law of physics of self-replicable molecules. Suppose we state Darwin's theory this way:

> Self-replicable molecules that are the fittest, adapt best to environmental stresses.

In this form, not only is there no tautology, there is no reference to survival. Can we generalize this law to a single living cell, to a conglomerate of living cells, to an organism? No, not yet. We cannot be sure that this form of Darwin's law can pass even the first hurdle, applicability to a living cell, let alone the rest of the hurdles. It is a long way from self-replicable molecules to a cell of many molecules with specific and distinct structures and functions, including the capacity of reproducing itself.

But we are getting ahead of ourselves. First we should ask, Can inanimate nature produce the kind of complex self-replicable molecule discussed above, to which the Darwinian mechanism applies? If there is no human intelligence around to guide their production, how could such molecules come about in the primordial environment of 3.8 billion years ago? Many biologists would be satisfied saying that such molecules could come about simply by chance. A few find such explanations wholly unsatisfying; they propose something quite different.

Holistic Models of Self-Organization of Life: So Far and Yet So Close

A few biologists who see the weakness of the chance-driven reductionist models have proposed an alternative mechanism, in which self-organization is accomplished with the help of special self-organizing fields that come into play in complex systems. Such scientists are sometimes called holists because they believe that for a composite system the whole is not just the sum of the parts.

There are several very talented holists dealing with the theory of the origin of life who would solve the materialists' "low probability" problem quite differently. These holists believe that life comes about because the universe is self-organizing (for a review, see Capra 1996). The low probability problem is overcome by the action of self-organizing fields that operate in highly complex systems.

Physics has many examples of self-organizing fields. For example, in a ferromagnet the individual iron atoms spontaneously arrange themselves into a magnet, each succumbing to the self-organizing field of the collective. Economists talk about the free market in the same way: If a sufficiently large number of businesses participate in the market, the invisible force of the free market takes over. In the same way, life is an example of autopoiesis, the self-creation of poetry through self-organization, declare the biologists Humberto Maturana and Francisco Varela (1980).

The physicist Ilya Prigogine (1980), who started this whole line of thinking and even got a Nobel prize for it, and the so-called chaos theorists who followed him (see Wesson 1991), saw in the dynamic structures of the previously discussed Belousov-Zhabotinsky reaction examples of order within chaos. The idea is that because the system is acting away from equilibrium, not all the energy is spent to increase disorder or entropy; some of it may actually be available to create new order, thus raising the probabilities for the emergence of life. To these theorists, life arising from chemical reactions in the prebiotic oceans of the Earth would be just a more complex example of order within chaos. In their thinking, Manfred Eigen's idea of the life-producing protein-DNA hypercycle is an extension of this kind of holism, and thus the low probability criticism is not necessarily valid.

As you can see, like the materialists' theories of the origin of life, the holists' ideas begin well with something concrete (the Belousov-Zhabotinsky reaction is as evocative as the Miller-Urey synthesis of amino acids) but then get bogged down in magic words. For materialists, the magic words are *chance* and *necessity*; for chaos theorists and their later relatives the complexity theorists, the magic word is *self-organization.*

Materialist reductionist ideas for the origin of life are refuted by probability calculations, contradictions, and paradoxes. The self-organization idea is harder to refute, but objections can be raised.

Can we get causal autonomy for such a self-organizing system? Chaos theory gives an appearance of free will because chaos dynamics is not easily predictable; at suitable junctures the system, under the spell of a so-called attractor, takes this path or that depending on the parameters. The path selections are enormously sensitive to changes in the parameters for certain ranges of their value. It is as if the system is choosing its own path of development. But alas! The freedom to choose is only appearance. Ultimately the system is determined by upward causation, as for any other Newtonian system. By way of contrast, Prigogine's theory is non-Newtonian, but in so being, it is not as reliable (or fashionable). Because chaos theory gives us orders within chaos that resemble the types of order around which Prigogine designed his theory, chaos theory has surpassed Prigogine's theory in popularity.

However, the order we get from chaotic systems may not be the same kind we are looking for in living systems. Consider an example of chaos-based dynamical self-organization and order within chaos: the formation of convection cells when a pot of water is heated on a gas stove. If you heat the pot gently, the heat flows upwards steadily via conduction. What happens when you turn up the heat? Now the water adjacent to the flame gets hot fast and, being less dense than the water on top, has a tendency to rise. But the weight of the water above has the opposite tendency, to keep it down. Eventually though, the hot water does break through in a rising plume, which is a convention current.

Now if the heating is done with careful control, the convention pattern is found to attain an orderly arrangement (order within chaos) of the shape of a honeycomb of hexagonal cells. We call this behavior "order within chaos" because the order is paid for by a huge flux of entropy (disorder) from the pot to the environment. Also crucial is the fact that the heat source at the bottom of the pot maintains the system away from thermal equilibrium.

Although many cases of order within chaos are encountered in nonliving nature and can be explained by chaos theory or theories of that kind, all such cases are of a simple structure. They are not complex on the same scale as the information-bearing structure of the DNA or RNA molecule.

It is sticky problems like this that keep chaos theory from serious consideration as an explanation of the origin of life. In frustration, we can only joke about it.

Three professionals are arguing about the world's oldest profession—no, not *that* one—the profession of God. The first
guy is a surgeon. He says, "God must be a surgeon. It takes a
very tricky operation to make Eve out of Adam's rib."

The second guy is an engineer; he says, "God is obviously
a chaos theorist. Otherwise how would He create the orderly
world from all that initial chaos?"

The third guy, a lawyer, says slyly, "But who created the
chaos?"

Another holistic approach is that of the complexity theorists
(obviously enough), who play the self-organization game through
computer simulation (Kaufman 2002). Here they have been successful in demonstrating that any network containing a large
enough number of components and connections or interactions
tends to spontaneously transform into a phase of organized behavior. Isn't this similar to iron atoms organizing themselves in a ferromagnet or businesses organizing themselves into a "free" market?

"Real" biologists who work with molecules and chemical reactions don't like computer modeling of life and criticize such work
as "fact-free science" (a phrase coined by the biologist Maynard
Smith). More seriously, the criticism invoked against chaos theory
applies here, too. Complexity theories can produce simple types of
order but not coded information as in the genetic code.

The physicist Paul Davies (1999), who has discussed this issue
in some detail, concludes that what is required is almost a complex
random number sequence of the base pairs (a base pair is the two
molecules forming a "rung" in the double helix of the DNA molecule; see figure 6b). However, only very specific random-number
sequences are suitable because the message has to be biologically
meaningful. The mystery of life is the simultaneous production of
both complex randomness and specificity.

The author William Dembski (2002) makes the same point
by saying that life is specified complexity. For example, the word

and is specified (i.e., it has a clear definition in logic) but not complex; a large random-number sequence is complex but not specified. But the random-number sequence denoting all the prime numbers between 2 and 101 (which was used in the movie *Contact* to discern an intelligent sender of microwave signals) is both complex and specified.

I submit that to produce both complexity and specificity we require both upward causation and downward causation. Upward causation is needed to give us randomness in the form of possibility waves that obey quantum probability calculus. Downward causation, via quantum collapse and conscious choice, is needed to give us specificity.

What are we to conclude from these notions of self-organization? Their historical importance lies in this: The line of thinking followed by chaos and complexity theorists has introduced one new feature into the equation of life—that of the self.

Materialists split up biological systems into a causal part (the genes) and a part that is controlled and caused (the form). As the philosopher Willis Harman puts it (Harman and Sahtouris 1998),

> There is a strong tendency in modern biology to assume that organisms have an "intelligent" controlling aspect and a passive, quasi-inanimate, controlled aspect. The controlling part embodies the essential principles of the biological state: the capacity to keep the parts working in relation to one another, to reproduce, to evolve, and to adapt. The controlled part is explainable strictly in terms of physics and chemistry. The organism is seen as a dualistic mechanism, comparable to a computer with its "intelligent" software and its passive hardware that responds to instructions coded in the program. (140)

In contrast, the idea of self-organization at least reminds us that an organism is in some well-defined way a "self," in the sense that it has an integrity that it maintains and reproduces. In other words, there is intelligence and "software" not only in the genetic part but also in the form, in principle.

Chaos theorists, and other thinkers before them (including the polymath Gregory Bateson [1980] and the mathematicians René Thom [1975], who generated a mathematical theory of form called *catastrophe* theory, and Benoit Mandelbrot, who gave us the famous fractal Mandelbrot sets), have wondered about how closely some of the forms in the living world resemble the mathematical forms generated by the chaos theory or catastrophe theory algorithms. Could this be a coincidence? I think not.

BEYOND SELF-ORGANIZATION: MATURANA'S THINKING

It was the biologist (more specifically, neurophysiologist) Humberto Maturana who first took the discussion about the nature of life a crucial step beyond self-organization. I first became aware of Maturana's work at a workshop given by the philosopher and consciousness researcher Heinz von Foerster. The program covered consciousness, the nature of the self, and related topics. In that workshop I first heard the word *autopoiesis*, a term coined by Maturana and central to his thinking, and I liked the concept.

Auto means "self," and *poiesis*, which comes from the same Greek root word as the word *poetry*, means "making" or "creating." So *autopoiesis* means "self-creating." First you must appreciate that self-creating is a little different from self-organizing. We use the word *self* in reference to systems like ferromagnetism or the hypercyclical reactions of Eigen, but can we really believe that there is a self—an entity that experiences—associated with these systems? Although Eigen's cycles and hypercycles are posited to generate a system that can self-organizationally self-reproduce and even evolve, we hesitate to call it alive. Does our hesitance arise because we suspect this system has no experiencing "self"?

In coining a new word for his ideas, Maturana was probing the issue of what makes something alive. Clearly "aliveness" requires more than a bunch of simultaneously occurring properties, the materialist definition. Clearly it cannot be as simple as a self-organizing system either. For Maturana, the crucial addition, the defining concept of life, is circularity. He said, "Living systems . . . [are] organized in a closed causal circular process that allows for evolutionary change in the way the circularity is maintained, but not for the loss of the circularity itself" (quoted in Capra 1996, 96). Components specifying the circular organization help produce and transform other components so as to maintain the overall circularity of the network. For a cell, the cell wall, the cytoplasm, and all the important components of the cell are now included in the network of circular organization that is a living cell as defined by Maturana.

There is more. Remember that I discovered Maturana at a workshop on consciousness. What's the connection to consciousness research? Maturana himself elaborates:

> Living systems are cognitive systems, and living as a process is a process of cognition. This statement is valid for all organisms, with and without a nervous system. (Quoted in Capra 1996, 97)

Cognition implies a cognizer that is consciousness. Similarly, circularity is an important component in the concept of self-reference (see chapter 5). In this way, Maturana came very close to an understanding of what life is. Working with his student and collaborator, neurophysiologist Francisco Varela, Maturana further developed his theory of life as a circular cognitive system creating itself continuously. Later, Fritjof Capra attempted to add more depth (Capra 1996).

Unfortunately, all these models of circularity and cognition are still materialist models. As I mentioned before, if we employ objects to create something, the something will still be an object. In the final reckoning, these models cannot provide any subject, a real cognizer, except for the vain hope that defines holism—that the whole is greater than the parts and so somehow arising from the whole there is a new entity, an autonomous self, a cognizer.

To understand life, and to include the full potency of the organism in the definition of life, we need more than a holistic philosophy. We need consciousness to be real and potent. When we include circularity in the definition of life and then place this definition within the primary reality of consciousness, we can arrive at a full understanding and complete definition of life. This step is the subject of the next two chapters.

7

The ORIGIN of the UNIVERSE, the ORIGIN of LIFE, and the QUANTUM OBSERVER

L et's go back a little, from the origin of life to the origin of the universe itself. The truth is, the two origins are intimately related. This is a subject that I have addressed more than once (Goswami 1993, 2000), and in this exposition I will follow the treatment given in my most recent book (Goswami 2008).

According to modern cosmology, a primordial explosion called the big bang created our universe. Good empirical evidence exists for such an explosive beginning about 15 billion years ago in the form of a "fossil" remnant of all the radiation emitted in the explosion, a microwave background radiation that now pervades our universe. Furthermore, the big bang explanation fits well with the fact that our universe is expanding, a fact predicted by Einstein's theory of general relativity and observed by the astronomer Edwin Hubble in 1928.

Einstein discovered the theory that explains the large-scale structure of the universe—the theory of general relativity, in which gravity is seen as the curvature of space-time. This theory makes room for the big bang, but only as a singular event. In the sixties, this concept acquired an exuberant following of God aficionados, including some astronomers: The singularity of the big bang must be the signature of divine creation! Alas, it is not that simple.

QUANTUM COSMOLOGY

Thinking of the origin of the universe as a creation event ex nihilo, from nothing, does not entirely satisfy, because even with a singular beginning we can always ask, What was before the singularity? Also, the presence of a singularity is not a particularly desirable aspect of the theory of general relativity. As the singularity is approached, some of the quantities of the theory, such as the energy density of the universe, tend to "blow up" in mathematical terms, signifying that the validity of the theory is questionable under those extreme conditions.

The physicist Stephen Hawking developed a quantum cosmology to avoid the singular beginning in time. There is no beginning; there is only possibility. The idea is that in the beginning, the cosmos must consist of quantum possibility. The universe must have begun as a superposition of many possible baby universes. But now we must ask, How does the superposition of possibilities become the actual universe in which we find ourselves?

Furthermore, consider the paradox that comes with pondering a universe of possibilities that can collapse to an actual event, the actual universe. It takes quantum consciousness acting through a sentient observer to collapse quantum possibilities. It is hard—no, impossible—to imagine that conscious observers were present during the hot early days of the cosmic big bang! What then?

How could the universe be here because of us, when we were not even there to greet it at the big bang? However, the universe could have been created in possibility in such a way that we would come into the picture in possibility and thereby bring the universe of possibility into manifestation. This mind-twisting idea is actually supported by several experiments, as we will see, but first we need to make one more stop in materialist territory.

A UNIVERSE OF CHANCE AND NECESSITY?

Many materialists think that we are here because of pure chance, some kind of cosmic accident, and that our consciousness evolved because of survival necessity. In materialists' thinking, there is no meaning anywhere in the universe and that includes us, our consciousness. "The more we understand the universe, the more it looks pointless," the Nobel laureate physicist Steven Weinberg said not so long ago.

The materialist model of the evolution of the universe goes like this. About a billion years after the big bang, after some expansion and some cooling, statistical chance fluctuations caused conglomerates called galaxies to condense. The galaxies themselves evolved, from spherical clouds of gases to disk shapes, many with spiral arms. Then stars began to condense, but these first-generation stars did not have all the elements needed to make life as we know it. However, after a few billion years, these first-generation stars "went supernova," self-destructing in an explosion that produced heavier elements. New second-generation stars condensed out of the debris of the supernova, some of them with planets. Some of these planets possessed both a solid core and a suitable atmosphere, just what's needed for the evolution of life.

The play of chance continued, claims the materialist. Statistical fluctuations and atmospheric energetics working together, completely by chance, made amino acids (the building blocks of proteins) or nucleotide molecules (the building blocks of the DNA and other nucleic acid molecules), or both. Proteins and

DNA are "living" molecules in some sense; they are the principal ingredients of a living cell, and they have a tendency to survive and maintain their form. Now, according to the scenario of materialist biologists, a new ingredient is added, the necessity for survival.

The picture above was initially supported by the Miller-Urey experiment, mentioned in chapter 6, which demonstrated that amino acid molecules do spontaneously form in a watery solution of the basic atoms (carbon, hydrogen, nitrogen, oxygen) if the energetics of the early terrestrial atmosphere is suitably simulated. But problems remained. The huge gap between the initial amino acid and the "living" macromolecule of protein has not been bridged. Nor has anyone ever been able to manufacture any of the full-scale nucleic acid molecules starting from the nucleotides. On top of this, theoretical calculations easily refute the idea that chance can assemble a macromolecule like a protein from its basic ingredients, the amino acids; the probability is much too small (Shapiro 1986). The biologist Michael Behe (1996) has cogently argued that the probabilities do not improve much even when we include survival necessity in the equation.

If life is not a product of chance and necessity, then is it a product of design? Is the universe purposive, made in such a way as to inevitably evolve sentience? Amazingly, today many astronomers and astrophysicists propound such an idea. It is called the *anthropic principle*.

THE ANTHROPIC PRINCIPLE: IS THE UNIVERSE MADE FOR MAKING US?

In its weak version (Barrow and Tipler 1986), the anthropic principle states that all physical and cosmological quantities take on values restricted by the requirement that there exist sites where carbon-based life can evolve and by the requirement that the universe be old enough for it to have already done so.

The strong version of the anthropic principle is even more emphatic about a relationship between the universe and the life in it. According to the strong anthropic principle (Barrow and Tipler 1986):

> The universe must have those properties which allow life to develop within it at some stage in its history.

Is the anthropic principle mere philosophy? No, much evidence for it exists, in the form of explanations for many weird coincidences. I will give you a couple of examples.

The universe is expanding over time, but if the force of gravity were even a wee bit stronger, the expansion would rapidly change into collapse, so there would never be enough time for life to evolve. If gravity were too weak, the universe would go on expanding but without any galaxies or stars to make a suitable environment for life. Furthermore, if the electrical force between electrons were even a little different, life as we know it would be impossible.

I could fill pages with such examples of purposive fine-tuning of the universe. One of my favorites involves the physics of the atomic nuclei—the physics of how three nuclei of helium atoms fuse together to make a nucleus of carbon, the crucial ingredient of carbon-based life. The conventional view of nuclear fusion reactions tells us that the probability of three helium nuclei fusing to form carbon would be too low to effectively generate much carbon in the universe. But guess what? The conventional view is wrong, because the frequency at which the three helium nuclei vibrate as they fuse together matches exactly one of the natural frequencies of vibration of the carbon nucleus. The effect of such frequency matching (called a *resonance*) is an enormous amplification of the reaction process, so that carbon forms far more profusely than we would expect. (Resonance can also amplify destruction: Soldiers marching in unison on a bridge can amplify a natural oscillation of the bridge so much that it collapses.)

How is it that the dance of three helium nuclei corresponds so exactly to one of the select repertoire of dances that six protons and six neutrons of the carbon nucleus can do? From a purely materialist point of view, the laws governing the nuclear structure of carbon are the result of random (called *stochastic*) motion of the material substratum. And these laws coincidentally conspire with the laws generated from the stochastic motion of the material substratum of the three helium nuclei? It truly boggles the mind. Isn't it much less mind-boggling, much more credible, to assume an intelligent designer stands behind it all? Isn't it less complicated to assume that there is a designer that designs the physics of both groups, designs the laws of physics of all nuclear movements, to make the resonance happen?

Further support for the anthropic principle came from the physicist John Wheeler (1975), who also lent quite a bit of support to arguments on the observer effect of quantum physics in its early days with his idea of the participatory universe. Wheeler thought about all the different observers in different parts of the universe manifesting all those different parts along with their properties . . . The result was much consternation on his part: Shouldn't the results of all those observations be chaotic? Or are the different observations somehow made consistent? He eventually proposed the idea of the participatory universe: Observers participate in order to make the universe consistent and orderly. Can you see how revolutionary Wheeler's idea is? How can an observer on a planet in the Andromeda galaxy choose so that the evolution of "her" universe is consistent with the evolution created by an observer on Earth, which is part of the Milky Way galaxy? If you think about it, this problem is the same as the paradox of Wigner's friend. If both observers get to choose in their individual consciousness, how can they possibly choose the same universe? The resolution, of course, is the same as the resolution of the paradox of

Wigner's friend: Observers choose from the vantage point of a cosmic nonlocal consciousness; that's how consistency is maintained. Of course, the great Wheeler himself missed the full implication of his idea of participatory universe: The universe is participatory via the downward causation of nonlocal consciousness.

Finally, Wheeler's articulation of a participatory anthropic principle (Wheeler 1975) can be summed up thus: Observers are necessary to bring the universe into being.

Of course, materialists can't let a challenge like this pass. Their answer to the anthropic principle is cosmological: Suppose our universe is one among many parallel ones all embedded in a "multiverse." What does this highly speculative proposal solve? If there are many, many universes, the odds must be better that one of them will be fine-tuned enough to produce life without the help of a designer. The odds are against that only if our universe is unique.

Well, this multiverse theory is not compelling for two reasons. First, even cosmologists admit that it is a highly speculative theory in a field that thrives on speculation. When cosmologists know how to verify the existence of other universes, then biologists can start thinking about the implications for evolution! Second, there's the issue of chance. I have already mentioned Michael Behe's intelligent design theory (Behe 1996). According to Behe, an "irreducible complexity" is built into life that makes it impossible to build life from matter via chance, multiverse or not. By using quantum physics, I have made this argument foolproof (see chapter 8).

The anthropic principle in any of the three versions mentioned suggests quite strongly that the universe is purposive, created by a designer with the purpose of creating life. Life and by implication we ourselves are here because of the way the universe is designed in possibility so we can collapse the possibility into manifestation. Then what about the paradox that we were not there when the big bang took place? Can our "later" choices affect things that happened "earlier"? Strangely, an experiment of quantum physics called the *delayed choice experiment* says the answer is yes and suggests the way out of the paradox.

THE DELAYED CHOICE EXPERIMENT

Again we encounter the ideas of the physicist John Wheeler (1977), who suggested an experiment to demonstrate that conscious choice, even delayed choice, is crucial in the shaping of manifest reality. This "delayed choice" experiment has been duly verified in the laboratory (Helmuth, Zajonc, and Walther 1986).

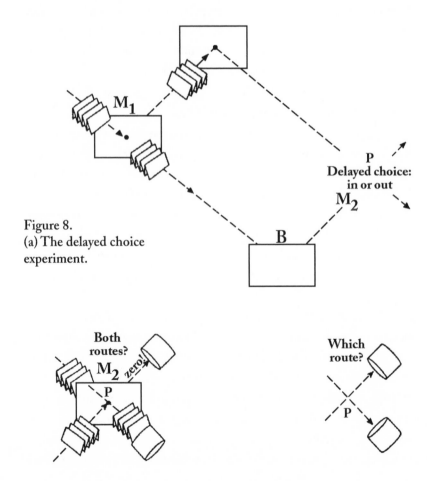

Figure 8.
(a) The delayed choice experiment.

In the delayed choice experiment, a light beam is split into two beams—a reflected beam and a transmitted beam—of equal intensity using a half-silvered mirror M_1 (fig. 8a). The two beams of light are made to cross each other at a point of crossing P using two regular mirrors as shown in figure 8a.

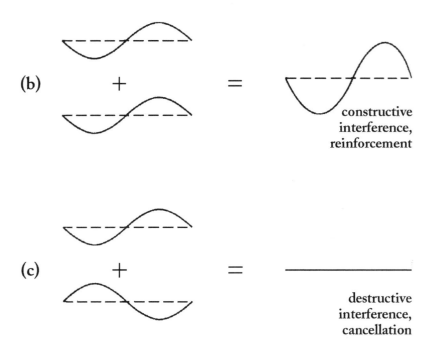

(b)   +   =   constructive interference, reinforcement

(c)   +   =   destructive interference, cancellation

Figure 8. (b) Waves appear at the same place in phase: constructive interference. (c) Waves appear at the same place out of phase: destructive interference.

If we put light (photon) detectors past the point of crossing as shown at the bottom right of figure 8a, one detector or the other will detect a photon fifty percent of the time. Each detection defines a localized path of the detected photon. We say that the photon shows its "particle" aspect in this experiment because its pathway is determined by the experimental arrangement.

But what happens if we put a second half-silvered mirror M_2 at P (fig. 8a, bottom left)? By splitting each of the two beams again to one reflected and one transmitted beam of equal intensity, the second half-silvered mirror forces not one but two beams to operate on each side of P (the transmitted part of one original beam plus the reflected part of the other original beam). Two beams operating together and recombining interfere with one another like waves; hence, with this arrangement you have an opportunity to verify the wave nature of photons. Indeed we can arrange the detectors in such a way that the two waves interfere constructively at the detector site on one side (fig. 8b) and destructively on the other side (fig. 8c). So on one side of mirror M_2 at P the wave is enhanced and the detector is certainly activated. But on the other side of P, the two waves come together out of phase and destroy each other (fig. 8a, bottom left). Here the detector gets nothing, and is never activated! How can this happen? To make sense of the experiment we must assert that the photons are no longer traveling in localized paths as before, one path at a time; they are traveling both paths until their detection, in possibility.

This experiment demonstrates that light is (as, indeed, all quantum objects are) both wave and particle. Each arrangement shows one facet of light; the facet we see depends on the arrangement we choose.

Wheeler suggested another twist in the experiment. A time gap of a few nanoseconds (a billionth of a second is called a *nanosecond*) occurs as light travels from M_1 to P. Suppose we insert the second half-silvered mirror at P during that time gap, after light has already been split at the first mirror but before it arrives at P. What happens now? If you think that the photons have already started on their designated path and will continue to show their particle nature, think again. The photons respond to even our "delayed" choice to put a second half-silvered mirror at P: They behave like waves and travel both paths.

On the other hand, if we are in the middle of the wave detection experiment and mirrors M_1 and M_2 are both in place in their respective location, what happens if we make the delayed choice to remove the mirror at P at the last nanosecond? Again the photons respond to our delayed choice and travel one path or the other.

Mind you, there is no puzzle or paradox here as soon as you reconcile in your mind that this is what it means to say that light is a wave of possibility! The entire path of the object stays in possibility until our observation manifests it retroactively. Yes, retroactively—going backward in time.

So it is with the universe: We manifest it retroactively. There is no manifest universe, only baby universes of possibilities, until we collapse it, until a sentient being appears in possibility in one of the possible baby universes and tangled hierarchically observes its environment. Only then.

The Delayed Choice Experiment in the Macroworld

The delayed choice experiment has helped many scientists change their attitude toward the observer effect and the import of the anthropic principle. But if you are a diehard, you may still be reluctant to accept the message of the experiment; you may say, "That experiment only applies to the microworld. I will believe the potency of the observer only when you demonstrate delayed choice in the macroworld that we inhabit." Well, the macroworld delayed choice experiment has been performed, and with flying colors, by the physicist and parapsychologist Helmut Schmidt and collaborators (1993).

Originally, Schmidt had been researching psychokinesis, the study of the movement of matter with conscious intentions. Over many years Schmidt had had a modicum of success in this enterprise. Some of his experiments involved random-number generators, machines that generate random sequences of zeros and ones starting with random radioactive decay events. Some people, through the power of their intentions, are able to influence the behavior of these machines, making them produce more zeros or ones according to their intentions, though only to a degree, of course. Incidentally, people with a power of intention this strong are often called psychics, but don't be put off by that word. In the quantum view, each one of us has this power; we just don't practice enough to use it effectively.

Schmidt's 1993 experiment is revolutionary because he combined the random-number generator experiment with the delayed choice experiment. In this experiment, the radioactive decay events took place, but no observer saw them. The detection of the decay product by electronic counters, the computer generation of random-number sequences, and the recording of information on floppy disks were all carried out without the information being seen by any human eyes. All this was done months before the psychic came into the picture. The computer made a printout of the sequence of numbers generated by the random-number generator, but, with the utmost care being taken to prevent observation, not only of the results but of the paper itself, the printout was sealed and sent to an independent participant.

The independent participant left the seals intact. A few months after receiving the sealed envelope containing the printout, this person instructed the psychic to try to influence the previously generated random numbers in a specific, randomly chosen direction (to produce either more zeros or more ones). The psychic then tried to influence the random-number sequence as directed. Only then did the independent participant open the seal of the envelope and check the printout to determine whether any deviation from randomness occurred in the direction indicated. If the psychic's efforts had no effect, the printout would be random.

A statistically significant psychokinetic effect was found in four of five independent experiments, each with a different independent participant. (The results were ambiguous in the fifth case.) Somehow, even by delayed intention, the psychics retroactively

influenced a macroscopic printout of data that in conventional wisdom had been made months earlier. The conclusion is
inescapable. No manifest thing in the universe exists until an
observer sees it. All objects remain in possibility, even macroscopic objects, until consciousness chooses from the possibilities
and an event of collapse occurs.

BACK TO TANGLED HIERARCHIES:
AN ANSWER TO "WHAT IS LIFE?"

The lesson of the delayed choice experiment is profound. It solves
the measurement problem of quantum cosmology—how the universe of possibility can be actualized even though no sentient
being was present to observe the big bang. The universe remains
as a superposition of baby universes that evolves in possibility
until, in one of the possible universes, the possibility of sentience
arises. Then quantum consciousness/God collapses the possibilities and the evolved first sentient being observes itself as separate
from its environment, whereupon simultaneously the universe
manifests retroactively, going backward in time from the moment
of collapse all the way to the big bang.

So it is true that we are here because of the universe and its
purposive design, but it is also true that the universe is here
because of us, our power of downward causation in our Godness.
There is circularity here, a breakdown of logic—quantum collapse
manifests not only the observed, but also the observer. This
dependent co-arising crucially depends on the circularity of the
logical chain, as discussed in the next section.

We must also ask the quintessential question, What qualifies
a biological being to be an observer? Are we human beings the
only observers in the universe? Does the universe of possibilities
wait in limbo until the first human observer comes into the scene
of possibility? This scenario conforms to the biblical creationist

idea that God created the universe some six thousand years ago. Well, almost.

From this point of view, the creationist idea is not nonsense, nor is it unscientific, as commonly assumed. Quantum physics shows us that much. But biologists need not worry. The idea conflicts with the fossil evidence. But couldn't the fossils have been created retroactively going backwards from the time of collapse from when Adam (in his God-consciousness) first *looked?* Unfortunately for Bible believers and creationists, this idea, too, contradicts the fossil data. Retroactive manifestation of fossils could only explain all the fossils that causally preceded the human lineage. The actual fossil record is much more diverse; it contains many other lineages beside the human lineage: the kingdoms of the plants and the fungi; the many other phyla of the animal kingdom besides the chordate phylum of which humans are a part; and so forth. Some of these lineages are even extinct lineages.

No, the combined lesson of quantum cosmology, the anthropic principle, the delayed choice experiment, and the fossil data are clear. Life itself, in the form of the first living cell, is the first observer. By realizing this, we also solve the problem of how to distinguish between life and nonlife in a succinct way.

In this way, it follows that there is no need to posit that tangled hierarchical quantum measurement can happen only in the brain. The concept of tangled hierarchical self-reference allows us to distinguish life and nonlife and opens us to a concrete definition of life.

The biologist Charles Birch (1999) wrote, "Biology right now awaits its Einstein for an answer to the question, what is life?" Well, Charles, the wait is over. Quantum physics comes to your rescue. If we ascribe to a living system the capacity for observership by virtue of tangled hierarchical quantum measurement, the biologists' consternation about defining life and at the same time distinguishing it from nonlife is over.

The biologist Humberto Maturana (1970) came close to this position by characterizing life as the capacity for cognition. Cognition requires a cognizer, thinking requires a thinker, perception requires a perceiver—and thus we come to the observer again.

THE OBSERVER AND CIRCULARITY

Behold the causal circularity of the role of the observer in quantum measurement. The observer, the subject, chooses the manifest state of the collapsed object(s); but without the manifest collapsed objects, including the observer, the experience of the subject does not arise either. This circular logic of the dependent co-arising of the subject and object(s) is called *tangled hierarchy*.

As discussed previously in chapter 5, the idea of tangled hierarchy and how it leads to self-reference, or the subject-object split of the world in manifestation, have been elucidated by the artificial intelligence researcher Doug Hofstadter (1980). So how does self-reference arise in the living cell? The answer: via tangled-hierarchical quantum measurement in its creation. And through this feat, accomplished in the first living cell, the universe itself became self-referent. The universe itself was able to split in two parts, one part of which (the biota) observes the other. Life introduces observership into the universe.

COINCIDENCES IN THE ORIGIN OF LIFE

If the biological problem of the origin of life is connected with the anthropic principle of the cosmos, it is reasonable to expect that, as with the anthropic principle, the origin of life must involve uncanny coincidences and evidence of fine-tuning. Indeed, an examination of the delicate situation demonstrates this quite amply:

- As Paul Davies (1999) points out, the origin of life requires at once near-perfect randomness and near-perfect specificity, a combination that requires fine-tuning.

- The probability of proteins arising by themselves is extremely small; the probability of the nucleic acids arising by themselves is also minuscule. But because probabilities multiply, the probability of the two arising together, nucleic acids with protein, is practically zero. When you consider on top of this the probability of the cell wall appearing, you cannot ignore the enormity of the coincidence involved. The enormity remains even when you consider the idea of a primitive, relatively uncomplicated nucleic acid and protein molecules confined by a primitive cell wall.

- Why are only twenty amino acids used in constructing life? Why not more and why not less? This specificity reflects extreme fine-tuning.

- The same comment applies to the genetic code. Why not some other code using some combination other than triplets of base pairs?

- Of molecules that occur in nature as both left-handed and right-handed forms, life uses only left-handed molecules. Why?

- In fact, most biologists agree that the arrangement of DNA is pretty nearly optimal, again indicating extreme fine-tuning. A little deviation from this optimal structure, and there might not have been any life!

Many more examples can be given, but too many technical details would be required. The conclusion is this: The situation for the origin of life may not be all that different from the situation of the cosmos. Purely material explanations work when the parameters involved have a lot of leeway, but when the parameters are very specific, finely tuned, purposiveness is implicit.

8

A DESIGNER
NEEDS BLUEPRINTS

*T*o the establishment biologist, any talk of design raises the specter of a biblical God; it just isn't politically correct. Quantum physics restores design talk to political correctness. The application of quantum physics to the question of life's origin, as presented in the last chapter, shows that the Genesis model of creation is wrong; it is, of all things, too close to classical physics. In Genesis, God created inanimate matter first, and then He created life. But the quantum God creates life *and* the universe in one fell swoop: The whole material universe waits in possibility until the first life manifests, and the self-referential quantum measurement circuit is completed via the experience of the first life.

Tangled hierarchical quantum measurement in the living cell requires both DNA and protein (and also, when you think about it, all the other basic ingredients of the cell). Thus everything must remain in possibility until this whole cell combination comes into the picture.

Could tangled hierarchical life have originated in stages, as proposed by some origin-of-life theorists? If there were stages, when the first primitive cell collapsed the universe and itself, all the earlier stages in life's production would collapse retroactively. We can ask, Is there evidence for stages? The fossil record of primitive life is very interesting. Primitive prokaryotic bacteria appeared about 3.8 billion years ago; blue-green algae, prokaryotic cells that existed both alone and in conglomerates, appeared starting about a billion years later. After another billion years, the first eukaryotic cells show themselves in the fossil record. Nothing in these records indicates that complex stages preceded the first prokaryotic cell.

The truth is, the step-by-step scenario is the creation of a Newtonian mind-set that cannot see beyond continuity. God is quantum consciousness; God works with creativity, and the operative concept is the discontinuous quantum leap. We could paraphrase Genesis with "God said, 'Let there be *life*'; and there was life," and we would almost be describing a quantum leap—except for a couple of caveats. First, even creative acts are preceded by a process; second, even if we say that life is the product of a creative act, we have to deal with the small probability question.

The effective way to deal with the question of small probability, judging from our own creative experiences, is to have the benefit of a blueprint. God's blueprints for the creation of life are called morphogenetic fields (see chapter 4).

A DESIGNER USES BLUEPRINTS: MORPHOGENETIC FIELDS

In considering God's creation of life, it's easiest to start with our own creativity. When we create, we are using God's power of downward causation through conscious observation. So how do *we* create?

Consider an architect designing a building. The architect starts with an idea from the archetypal domain. At the second stage, the architect makes a blueprint of the idea. Only then will he or she participate in actually creating the physical building by putting the physical ingredients together.

Quantum consciousness/God follows the same procedure in creating life. He/She/It starts with the possibility of an idea that belongs to the supramental domain. And only then does She/He/It get busy looking at the domain of possibilities having to do with the blueprint of life, the domain called the vital energy body, while simultaneously looking among the material possibilities for a physical representation that fits the vital possibilities.

Now let's go back to the beginning of life. The probability of making proteins and DNA individually is minuscule; this we know. We also know that putting DNA and proteins together through cycles and hypercycles of chemical reactions is also highly unlikely—where does it begin? But do we need a continuous process getting to the end product? Let's invoke the discontinuity of creativity for the completion of the tangled hierarchical quantum collapse in the first living cell.

The big hint here is the already tangled relationship between proteins and DNA. Each is required to make the other; they are co-creators. The designer, quantum consciousness/God, recognizes the tangled hierarchical situation of the protein-DNA combination in possibility, because He has the blueprint of a living cell, also in possibility, with which to compare it. The blueprint codifies the knowledge that to create a self-reproducing living cell, one needs a replicator system (which is physically represented as DNA with genes as components) and maintenance managers (physically represented by the proteins and enzymes) and a cell wall for confinement, etc.

The mathematician John von Neumann (1966), working with bits of programmed material called cellular automata, figured out theoretically the roles of the replicator and the maintainer systems in the production of a truly reproductive system even before Crick and Watson discovered these roles in the laboratory.

In realizing that both the replicator and the maintenance manager are program-like, von Neumann went one better than Darwinians, who almost universally assume that only DNA is program-like, that the only living software is in an organism's DNA. That is not true. The proteins, cells, organs, and entire organisms all are designed to perform purposive programmed functions under conscious guidance with the help of the nonlocal morphogenetic fields.

However, von Neumann missed the importance of considering the self-referential capacity of living beings, so he missed the tangled hierarchical relationship between DNA and proteins in their making. Furthermore, he did not anticipate the necessity of invoking God-consciousness in the origin of life. Only because we have recognized the importance of self-reference are we now able to figure out why God is needed for the creation of life and how God must have created the first life.

If we insist on positing downward causation emerging in a self-organizing system from holism, that is, because the whole is greater than its parts, we implicitly assume an organizing principle apart from what is inherent in matter. The biologist Michael Behe (1996) intuits intelligent design, and he introduces the concept of "irreducible complexity." By this term he means biological features that are so complex that they implicitly require an intelligent designer—God—to design them. Why not make the organizing principle explicit?

The quantum consciousness/God hypothesis makes the magic explicit. The magic of life comes from God's creative downward causation, and it's really not magic. It is the same downward causation of insight that is involved in any quantum collapse event of creativity. The insight simultaneously manifests the vital-body blueprint of a living cell and the physical representation, namely, the following: (1) proteins for enacting the programs of living and amino acid sequences arranged in a polymer chain to generate the proteins; (2) DNA, with the various genes whose base pairs can code for the amino acid sequences to be assembled for the various proteins; (3) RNA for mediating the transfer of genetic information and to provide templates for building the amino acid sequences; (4) cytoplasm containing the building blocks of proteins and the nucleic acids and all other material (such as the energy-producing molecule adenosine triphosphate [ATP]); and (5) a cell wall for the confinement of all the above. This creative insight and its vital and physical representations simultaneously lead to the choice and collapse of the manifest first living cell—life. The tangled hierarchy in the creation of the living cell (which is most apparent for the DNA-protein duo) assures the self-referential quantum collapse in a living cell, and at once allows the cell to become an individual entity with apparent separateness from its environment. The magic in the creation of life is the discontinuous quantum leap of the creative insight and simultaneous representation-making; it can never be broken down to bit-by-bit continuous evolution by mathematics or mechanical demonstration.

Paul Davies (1999) was perturbed that the problem of the creation of life is the problem of creating at once a random sequence of nucleotide bases that are also specific and meaningful in their ability to code proteins. For creative nonlocal consciousness, this is not a step-by-step process. The proteins are part of the phenotype of the cell, the observed characteristics at the macrolevel of the genes. Quantum consciousness, while unconsciously processing the possibilities, processes the phenotypes (including the proteins) in potentia, that is, within the plane of quantum possibilities. When a fit is found between the blueprints of life in the vital-body domain and the cell (including the proteins) that represents them in the physical domain, a choice is made. The collapse of the cell (including the proteins) collapses automatically the genes that causally must precede the proteins. All the components of the cell remain in possibility until consciousness chooses to actualize them. In the process of actualization, consciousness identifies with the autopoietic organization, the living cell.

This identification of consciousness with the living cell is the beginning of an evolving self. The evolution of life is also the evolution of this self or individuality. For a primitive prokaryotic cell, it is the tangled hierarchy (as in the DNA-protein combination) in the creation of the cell that gives rise to the self. And each process of reproduction (in this case, just simple cell division) involves once again the tangled hierarchical creation of the DNA-protein duo and thus a new identity of consciousness with a new cell. We call this repetition the life cycle.

As more complex organisms evolve, the life cycle consists of, metaphorically speaking, an acorn becoming an oak to produce more acorns, producing more oaks. Acorn or oak, which came first? This circularity is the reminder of the basic tangled hierarchy of life.

MENTAL AND VITAL CREATIVITY

The magic of biological creativity confuses us more than our ordinary mental creativity, which is confusing enough. Many people still have difficulty accepting the discontinuity of ordinary mental creativity—creativity as discovery and invention of new meaning. But creativity in the vital plane involves a yet deeper degree of discontinuity.

Mental creativity involves a discontinuous insight in the movement of the mind. It requires a tangled hierarchical quantum measurement device, namely, the brain. However, when we make a physical representation (a product in the outer arena) of our insight, we can do it bit by bit, continuously. No discontinuity is involved in this manifestation stage.

In creativity of the vital plane, consciousness discontinuously (in the vital domain) discovers the vital blueprint and simultaneously in the physical plane makes a physical representation of it;

the physical representation also acts as the quantum measurement device to facilitate collapse. Hence, the physical representation has to involve a tangled hierarchy, and therefore its creation also involves a discontinuity.

The Definition of Life

We are finally in a position to give a complete definition of life:

> A living being consists of tangled hierarchical quantum measurement apparatuses that are representations of the vital blueprints of biological functions including but not restricted to maintenance and reproduction. Such a being is capable of self-reference because in the process of quantum collapse involving it, consciousness identifies with the being.

This definition has an added advantage for complex organisms: It helps explain not only the complexity of living, but the complexity of dying. For a complex organism, not only life but also death is more complicated. Beyond the basic level of cell death common to all organisms, a complex organism experiences higher levels of death, corresponding to cessation of organ functioning. In other words, there is cell dead, and there is organ dead.

Animals with a brain have yet another complication, the highest level of death: brain death. In fact, for us humans, when someone's brain is dead, we would tend to call the person dead, period. The reason is that, in addition to making representations of vital blueprints, the brain makes representations of the mind, a "higher," more sophisticated facility with which we identify more completely. So when the capacity of making representations of the mind ceases, we are considered dead.

CHAPTER EIGHT

THE CHALLENGE
TO MATERIALISTS

In the eighteenth century, the biologist Georg Stahl made the following challenges to the purveyors of materialist biology, who looked at life in purely mechanical terms:

1. It is impossible to make organic molecules from inorganic ones.

2. Life is not mechanistic physical machinery but transmutational.

3. Living organisms cannot be machines because they suffer.

To this challenge the biologists quickly adjusted. They responded by agreeing with the second point and finding a chemical, instead of a physical-mechanical, basis for life: Life is chemical transmutation of substances. For the first point, they hailed the synthesis of the organic molecule urea starting from inorganic components and assumed that the issue was settled. As for point three, biologists en masse ignored it.

The present theory of life offers the following challenges to the materialist:

1. It is impossible to manufacture through continuous step-by-step processes tangled hierarchical devices that are essential for life starting with simple hierarchical systems. Tangled hierarchical quantum measurement is how a living being acquires conscious identity, that is, subjecthood or observership.

2. Life is not only transmutational, with a chemical basis, but also representational; that is, the chemical machinery becomes *bio*chemical as it makes representations of the subtle blueprints purposive of biological functions.

3. Living organisms, because they have feelings, cannot be reduced to inanimate machines.

You can see that even in the eighteenth century, Stahl was on the right track; he just didn't have access to the modern concepts needed to make his challenges stick. Still, he got one criterion of why life must be different from nonlife—the criterion of feeling—completely right. We'll look more at the importance of feeling later in the book (see chapter 14).

SOMATIC PROGRAMS
IN A LIVING CELL

I contend above that there are programs not only in the genes (the genetic code and the programs that run the regulator genes) but also in the somatic part of the cell. In contrast, establishment biologists always tend to assume that the genes have all the programs for cell functioning. The cell biologist Bruce Lipton's (2005) work supports my contention. Says Lipton:

> I had been trained as a nucleus-centered biologist as surely as Copernicus was trained as an Earth-centered astronomer, so it was with a jolt that I realized that the gene-containing nucleus does not program the cell. Data is entered into the cell/computer via the [cell] membrane's receptors [special proteins embedded in the membrane], which represent the cell's "keyboard." Receptors trigger the membrane's effector [action] proteins, which act as the cell/computer's "Central Processing Unit" (CPU). The CPU effector proteins convert environmental information into the behavioral language in biology. (92)

Although he's moving in the right direction, Lipton's model for somatic programs shouldn't be taken too seriously, because his explanations are still materialistic. Furthermore, Lipton believes that even in the single eukaryotic cell, the organelles already express most biological functions. The right way to think, in my view, is that in advanced organisms, the organs are the software representations or expressions of vital blueprints and eventually of biological functions.

PART 3

CREATIVE EVOLUTION

CHAPTER

9

The €VOLUTION *and* 𝓕UTURE *of* 𝒟ARWINISM

s a physicist I have great reverence for Newton, as most physicists do. But when I write something about Newton's physics for the public, I don't feel compelled to say something nice about Newton or his theoretical contributions. The same is true, in general, of most physicists.

The manner in which biologists treat Darwin is entirely different. Virtually every popular reference to Darwin and his theory is accompanied by superlatives. Biology has been guided by Darwin's theory for the better part of a hundred and fifty years, so it is not surprising to find biologists reverential toward Charles Darwin. But why the compulsive praise?

I think this attitude has more to do with the politics of evolution than anything else. As a rule, biologists try to convince the general public of the importance of Darwin and his theory every chance they get. This tactic is pretty harmless. But this same politics-driven attitude promotes another attitude that is not so harmless: When it comes to seriously criticizing the scientific merit of Darwin's ideas, and especially when proposing a major revision or upheaval, biologists run scared.

A clue to their fear can be found in a telling comment by the biologist Julian Huxley in an article published in 1960: "The first point to make about Darwin's theory is that it is no longer a theory, but a fact." A fact based on watertight data? No, not then, and not now. The main source of data for evolution is the fossil record, and it has gaps. The situation has improved a little since the 1960s but not much. In Huxley's comment, "fact" means belief, a dogma. A dogma cannot be revised or modified with impunity; a dogma can only be interpreted. With Darwin's theory, we find a plethora of interpretations. So biologists who dare to challenge the theory get scared: They're taking on a dogma.

How did it get this way? How did a scientific theory that should be spoken for (or against) only in the context of experimental data or in terms of internal consistency take on a significance bordering on religious dogma? I'll examine that question in this chapter by putting the theory in historical context, including revisiting some aspects already discussed, as a way to set the perspective for the following three chapters, which propose a new direction for the future.

THE BEGINNINGS

Charles Darwin did not set out to promote an antireligious dogma. He was an immensely creative scientist whose great question was this: How do organisms manage to adapt themselves so well to their environment? Darwin actually began as a student of theology, so he must have been familiar with the biblical theory that God created all life, all at once, circa 4000 BC. But his interest in science must have propelled him to the awareness that geological time extends far beyond six thousand years. Through this immense geological time, Earth's environment was bound to have undergone immense changes. To remain adapted to the environment, it followed then that the organisms must also have undergone immense change. In other words, organisms must evolve, or so Darwin must have intuited.

Like a true scientist, Darwin went on to collect data and on the basis of that data became convinced that we evolve from ancestors. A species originates in some other species that originated from yet another species before it. The introduction of this concept marks, as the biologist Ernst Mayr puts it, the first Darwinian revolution.

But Christian theology does not have scope for such thinking. Darwin become disillusioned with Christian theology and began looking for a materialist motif for evolution.

For Darwin, the central themes of evolution are twofold: Species change, and species adapt to their environment. His insight was that change must come to a species through the accumulation of small random variations in some as yet unknown hereditary component of organisms. This was a highly original and ingenious idea. Mind you, the major hereditary component of life, the genes, had not been discovered yet. A theory of heredity existed, sort of, but it was based on the notion of inheritance based on use and disuse of organisms' acquired characteristics (an idea credited to Lamarck). In fact, Darwin himself, in the last paragraph of *On the Origin of Species*, seems to subscribe in part to the Lamarckian idea of heredity because the phrase "use and disuse" appears in his explication of variability. But by and large in *The Origin of Species* he left the question of heredity as a black box (not a bad strategy when beginning a new kind of science). Subsequently, Darwin tried to develop a theory of heredity, without success. Eventually, of course, the problem was solved by botanist Gregor Mendel's discovery of the gene.

But what about adaptation? The idea of natural selection may have come to Darwin in part from his own observation of animal breeding—the breeders select the trait they desire to breed. Darwin also noticed that organisms produce more offspring than can survive. Naturalist Edward Blyth had observed that nature tends to weed out the unfit individuals to help the species as a whole to survive. But creativity researchers who have studied Darwin's work also marvel (see Gruber 1981) at the emotional intensity Darwin experienced when he read Malthus' work on overpopulation and the competition for survival. If theology does not deal with the scientific question of adaptation, why not try materialism?

Thus originated the idea of natural selection: Inanimate and objective nature selects those variations that are adaptive to the environment. Also, whereas Blyth saw natural selection as a mechanism for species homeostasis, Darwin saw environmental adaptation as part of a more general process of change in the living world—evolution.

Darwin presented a two-pronged theory of the mechanism of evolution: (1) variation (driven by chance) in the hereditary component of living beings and (2) natural selection (driven by the necessity of survival), from among these variations, of the fittest to survive (the struggle for existence). Again, according to Ernst Mayr, this notion initiated the second Darwinian revolution in biological thought.

Despite their revolutionary power, the proposed mechanisms have their problems. The generality of the idea of natural selection makes it very difficult to falsify, a concern we will return to later in this chapter. And of course, the manner in which Darwin introduced the idea of natural selection contains a tautology, as I noted in chapter 6: Who is the fittest? One who survives. Who survives? One who is the fittest. Only very recently have we learned to define *fit* in such a way as to rescue Darwinism from tautology (see chapter 6).

Darwin was also quite aware that his gradual mechanism might not be adequate to explain such gigantic leaps of evolution as the emergence of the eye, which he worried could evolve only in an all-or-nothing way. Nevertheless he stuck to his guns, again, I think, because he could not see any alternative to gradualism.

Finally, Darwin was respectful of the purposiveness of biological beings and worried whether it was entirely correct to exclude purposiveness from biology, as his theory of evolution did. Toward the end of his life, he wrote lamentingly, "My Theology is a simple muddle. . . . I cannot look at the Universe as the result of blind chance, yet I see no evidence of beneficent design in the details." He also wrote, "I cannot see as plainly as others do . . . evidence of design and beneficence on all sides of us. There seems to me too much misery in the world. I cannot persuade myself that a beneficent and omnipotent God would have designedly created the *Ichneumonidae* with the express intention of their feeding within the living bodies of caterpillars or that a cat should play with mice." It never occurred to Darwin that it takes an evolution of consciousness to arrive at the ethical and virtuous behavior that we see occasionally in humans and that such evolution could never arise from chance-and-necessity-driven matter.

In short, Darwin's theory of evolution of life is a perfectly good piece of research in a developing field, nothing more. We can see that Darwin was embracing materialist metaphysics only reluctantly, because he could not see any other way to explain adaptation. But then people got to thinking that the ideology is the thing and the science and the data were secondary to it. This mindset led interpreters and extenders of Darwinism to muddy the situation even further.

A chief source of difficulty is that materialists saw in Darwin's theory the equivalent of Newton's paradigm of mechanistic and deterministic physics. Newton, through his laws of motion, established what defines stasis for a physical system: the state of rest or of uniform motion. He also established, and this tenet is most important, how a physical system undergoes change—through the action of an external force. The parallel of Darwin's theory to Newton's is obvious. For Darwin's theory, the species defines the state of homeostasis in biology. And the two-pronged action of variation and natural selection, chance and survival necessity, defines the mechanism for change of a species into another, or speciation.

The parallels are not completely satisfactory, however, and the discrepancy has been the subject of some controversy. When we apply Newton's laws to the movement of bodies, the bodies themselves, while undergoing movement, do not change. But the species does not remain entirely unaffected by the forces of change, the variation-and-selection mechanism. It is a difficult question of philosophy to decide when a particular homeostasis of a particular species can be said to have come to an end and a new species to have begun, because a species is continuously changing because of the variations in the hereditary material. The history of Darwinism and biology is filled with such debates.

Aside from such differences, the commonalities between Newton and Darwin have greatly shaped many of the intellectual habits of biology as a discipline. For both Newton and Darwin, change is continuous and gradual. Hence, as a discipline, biology has a considerable tendency to suppress ideas of discontinuity. Change is local in Newtonian physics, guided through interactions in the immediate vicinity through the exchange of energy via local signals. So nonlocality is a taboo in biological theory as well. Finally, like Newton's physics, Darwin's biology was also destined to be deterministic. If a Newtonian kind of deterministic mechanism extended to living systems, as Darwin's theory seemed to have established, then the entire natural world, the nonliving and the living, would be deterministic.

In the eighteenth century, the physicist Pierre-Simon, marquis de Laplace wrote a book on celestial mechanics that made no mention of God, practically a heresy. So Laplace was summoned by the emperor Napoleon, who promptly inquired, "Monsieur Laplace, you have not mentioned God in your book." (Even Newton did, by the way.) To the emperor's question, Laplace's famous answer was, "Your majesty, I have not needed that particular hypothesis." That took care of the movement of the nonliving universe. Darwin's work removed God from the living universe as well.

Let's remind ourselves of the views of that era. The prevailing philosophy in Darwin's time was modernism, a two-domain approach to reality. The external domain of matter was recognized as the domain of science; the internal mind was the domain of human free will and God's intervention (downward causation). This separation was the Cartesian truce between science and theology that enabled science to free itself from the shackles of religious dogma.

In Descartes' thinking, animals were machines and determined (a conclusion required by Christian beliefs, actually), but humans had free will and were beyond mechanism. Darwin, by proposing animal ancestry to humans, took the next step and proposed to demolish the Cartesian idea of human free will, instead pronouncing human beings determined as well. If humans are determined beings, then there is no room for their free will to respond to God's downward causation, no room for God's intervention. God is rendered impotent.

In this view, it is paramount that organisms, including the human organism, be determined machines. Thus, a three-pronged ideological attack began. The first prong was against purposiveness in biology, including biological evolution. Darwin himself was quite open to accepting purposiveness, and data in its favor seem self-evident. However, *determinism* means determined by "material and initial causes." Purpose smacks of final causes (sometimes called *teleology*); such causes are not compatible with materialist determinism. Hence the crusade. The second prong was directed against Lamarckism and, by association, against organismic ideas in general. Such ideas give importance to the organism and open biology to causation at the level of the organism, such as autonomy and free will, which cannot be reduced to materialist upward causation. The third prong goes beyond Darwinism and evolution to question the nature of life itself, of that which is evolving. If materialism is to be a complete success, it is impera-tive that life itself must consist of an epiphenomenon of matter and mechanism without any reference to anything nonphysical or, more specifically, anything from the domain of vital energy.

These three prongs of attack became the long-drawn-out crusades of Darwinism (which more and more became synony-mous with evolution) against teleology and purposiveness, against Lamarckism and organismic theories, and against vitalism. These crusades were all ideologically motivated and continue the same way; hence the venom seen even today on these topics, a tone of debate that scares the radical biologists (who do exist) to no end. In what follows, I will discuss each of these prongs individually.

THE CRUSADE AGAINST LAMARCKISM

The crusade against Lamarckism developed in three main stages. At the first stage, in the 1880s, the biologist August Weisman (1893) was the first to suggest in a concrete manner that informa-tion can only flow one way, from the hereditary (micro)material, which he called the *germ line*, to the soma, but not the other way around. At the second stage, when neo-Darwinism was formulat-ed (a new synthesis integrating genetics and population biology with Darwinism), Weisman's doctrine was reasserted using the concept of the gene. Then in the 1950s, when the structure of DNA was discovered, Francis Crick elevated Weisman's idea to the "central dogma of molecular biology" and now based it on the language of molecules: Information can flow from DNA to pro-tein and not vice versa.

Let's briefly ask why it is so necessary to assert this idea of one-way flow of information from DNA to protein, to make it into dogma. You see, reductionism is a fundamental aspect of materialism: Micro makes up macro, and causality moves up the ladder from micro to macro. So the *causal* flow of information is upward. Here we discover the even deeper dogma of upward causation. So genotype (micro) affects phenotype (macro); this arrangement is commensurate with reductionism and upward causation. But if the phenotype traits (the result of proteins, mainly) could affect the genes, that would be the macrolevel affecting the microlevel, completely against the grain of reductionism.

What the materialists are missing is, of course, the idea that genes do not have to be the only way for information to pass from generation to generation. We do inherit more than the genes from the previous generation, egg cytoplasm at one extreme and culture at the other extreme being two examples that even the materialist would allow. Accepting this broader definition of inheritance can open new possibilities for a neo-Lamarckism that biologists themselves have begun investigating (Jablonsky and Lamb 2005). When we do biology under the primacy of consciousness, when we permit the living the advantage of the vital body morphogenetic fields, a new avenue opens for Lamarckism to operate. All this is the subject of chapter 20.

The Crusade against Vitalism

The crusade against vitalism is even more important for materialists than the crusade against Lamarckism because vitalism is a threat to the most fundamental tenet of materialist philosophy—the primacy of matter. If life requires a vital substance to operate, then where is the supremacy of the material?

For Darwinists and evolutionary biologists, vitalism poses an additional threat because it can justify an evolutionary thrust toward more and more complex manifestations of purposiveness. Darwinists are all too aware that the biological arrow of time, the progression of evolution toward more and more complex life, must be reckoned with. If data exist, along with a theory for explaining the data, the situation could prove too embarrassing and too difficult to rationalize away. Hence, mainstream biologists have always opposed vitalism.

Initially the opposition was based on a two-fold approach. First, vitalism was seen as dualism. The question is posed, How does a dual nonmaterial vital substance interact with a material substance? This query puts the vitalist on the defensive. Second, anything vital that we may feel is considered part of the interior subjective experience. The idea of epiphenomenalism was posited: All our internal experiences are epiphenomena of matter, of the body.

And, of course, when molecular biology was solidly established and the functioning of the life of the cell began to be elucidated in unforeseen detail, vitalism just faded away. Eventually, it became every biologist's prerogative to deny any vitalistic tendencies.

However, as I mentioned earlier, vitalism made something of a comeback in the 1980s with Rupert Sheldrake's (1981) introduction of the idea of nonlocal and nonphysical morphogenetic fields. Morphogenesis is one of the weak underbellies of the materialist body of literature. The objection of dualism could be raised, but quantum physics can remove this objection, as discussed in chapter 4.

DARWINISM AND PURPOSIVENESS

I have discussed purposiveness to some extent already in connection with law-like and program-like behavior (see chapter 4). Here I will give the broader context for that discussion.

I said before that purpose is self-evident for biological systems. Suppose you have to explain the liver to a child. You say the liver helps maintain the body through metabolism. This is a purposive statement; you speak of the function of the liver—maintenance—in a purposive manner. Biological systems such as organs are purposive organs, and so, naturally, talking about them without referring to their purposive function is not very useful.

A materialist would say that we are being too hasty in concluding from such statements that purposiveness is a compulsory attribute of biology. Suppose we can express the same sentence without the purposive language. For example, we can say the liver secretes bile that helps digest the food we eat. This statement mostly avoids the taint of purposiveness. Materialists claim that it is possible to state biological propositions about organs in a purpose-free fashion, just as we state propositions in physics. Materialists further claim that it is the use of purposive language that confuses us about the importance of purposiveness and teleology in biology. The purposiveness is not real, only appearance; it is teleonomy, not teleology.

However, biological materialists can't really point to the example of physics any more, because it is no longer true to say that physics can be expressed in a language devoid of purpose. Take the anthropic principle—the universe is so constructed as to make life possible. That is a purposive statement. When a quantum physicist says that quantum mathematics generates superpositions of possibilities (possibility waves) for consciousness to choose from, he or she is also making a purposive statement. In the new paradigm, science within the primacy of consciousness, we can expect purposiveness even in the language of physics.

In truth, though, it is not an issue of language at all. It is an issue of our beliefs about biology. If we believe that biological beings, including organs and organisms, are run by DNA programs, then all the purposive behavior of organs and organisms would be an appearance only. For such apparent purposiveness biologists use the word *teleonomy*. As for the origin of the DNA programs, if we believe they arose by pure chance and evolved through chance and necessity, we have eliminated teleology and purposiveness from biology. But have we?

Just remind yourself of the following three things: (1) the DNA programs (the code for making proteins and gene regulation for making form) could not have arisen by chance (see chapter 6);

(2) making functional forms involves nonlocality and therefore could not be carried out by DNA programs alone (see chapter 4); and (3) Darwin's theory, though designed to eliminate purpose, is itself wrapped up in a purposive context of survival.

Face it, purposiveness cannot be eliminated from biology. If we take the self-evident purposiveness of the organs as part of our bio-philosophy, then we recognize the program-like nature of biological beings from the outset. Then naturally we attempt to understand their program-like nature with the help of the concept of a programmer and blueprints for the programs (see chapter 4).

Let's get back to the reason for the attack against purposiveness, namely, that in Darwin's theory evolution is not goal directed, not purposive. If evolution is purposive then Darwinism is wrong. On the other hand, evidence suggests that evolution does occur in the direction of increasing complexity that results in a hierarchy of being—a clearly goal-directed outcome. Some kind of adjustment is called for.

Recently, biologists and bio-philosophers have changed their focus from correcting purposive biological language use to co-opting some of the indisputable evidence of purposiveness in evolution and making it part of Darwinism. Such evidence is not hard to find. Humans have at least the capacity of self-awareness— we know that we are aware. We also have the capacity for processing meaning and purpose—witness this discussion. We occasionally have bouts of altruism, and some of us even try to live an ethical life. Where do these attributes come from? For the materialist, the source has to be Darwinian evolution, where else?

So Darwin's followers have been busy developing plausible theories of how ethics in our behavior may come about, why we evolve self-awareness and meaning-processing capacity, and so on. They have proceeded by interesting steps. The first move: invoking the idea of the selfish gene to explain away ethics as a consequence of genetic determinism (Dawkins 1976; Ridley 1996). The next move, even more ambitious: explaining away meaning processing, self-consciousness, and even the appreciation of beauty, as Darwinian adaptation (Dennett 1995). The final step, and the most interesting: claiming that Darwinian adaptive evolution itself brings about the capacity of evolvability, the ability to evolve in a canalized manner (i.e., following a preferred direction, as in a channel) toward increasing complexity (Kirschner and Gerhart 2005).

All this maneuvering is quite in line with a general tendency that has characterized Darwinists from the beginning. In the Middle Ages, when Ptolemy's Earth-centric theory of the world began to show disagreement with the growing observational data in astronomy, adherents of the Ptolemy paradigm busily invented a seemingly endless series of cycles and epicycles (circles within circles) to account for the movement of heavenly objects around the Earth, tweaks that allowed them to continue to justify the old paradigm. The same thing happened and continues to happen in biology. The Darwinists' response to any possible observational discrepancy is to propose a suitable modification of Darwinian ideas—shades of cycles and epicycles. Darwinism is so general that it can be reinterpreted to incorporate any data that contradicts it. It is not falsifiable.

The "Cycles and Epicycles" of Darwin's Theory

Without a doubt, Darwin's theory has had an enormous influence on subsequent scientific research in biology, which is one of the major goals of a scientific theory. With the development of genetics and population biology, a much-needed synthesis was accomplished of the old Darwinian ideas and the new developments. This synthesis, called neo-Darwinism, has somehow become dogma. To many biologists today, Darwinism, or rather neo-Darwinism, has become synonymous with evolution.

Can we say that Darwin's theory has been verified? Well, some aspects of it certainly have been. The idea that evolution happens, that species evolve from ancestor species is, I think, established beyond doubt, much to the chagrin of creationists and intelligent design proponents, who believe otherwise. The hereditary component of organisms that Darwin theorized about has subsequently been found—the gene—and much is now known about it. There is also abundant evidence of species fully adapted to their environment. So, like evolution, adaptation of species to their environment undoubtedly occurs.

For microevolution of bacteria, a few cases have been documented in which the whole package of a Darwinian gradual mechanism for speciation—variation and natural selection—seems to work, if we define speciation as reproductive isolation, as is now the fashion. In a few cases, such as the horse, one aspect of Darwinism—the evolutionary adaptation of a species to environmental changes in an ongoing way through gradual changes—has been verified to some extent.

Can we say, on the other hand, that Darwinism has been falsified? Certainly there remains much scope for discontent. Some of the discontent is based on what are perhaps minor matters. For example, we all know that animals have mobility, and so they can choose their own environment, at least in principle. When animals can leave the environment in which they were born, does it make sense to say that animals *must* adapt to their particular native environment? The issue becomes more intriguing when you ponder the creatures that migrate seasonally from one environment to another. So why don't all species migrate until they find the environment that suits them?

Adaptation raises other questions. Many examples can be cited of species with nonadapted components, including those that undershoot and overshoot the ideal adapted goal. How can that be? Still more disturbing are the many cases of species having developed an adapted feature before the environmental change actually takes place. How can that be, if the environmental change drives evolutionary adaptation?

Then there's the issue of speciation itself. The title of Darwin's seminal book is *On the Origin of Species*. But "species," as I have already mentioned, is a very ill-defined concept when change is presumed to occur more or less continuously. Under such change, when can we say that speciation has taken place? Darwinians say that when the evolving organism has produced two lines that cannot reproduce with each other, we must say that speciation has occurred. But given that most organisms (except at the microlevel of bacteria, for example) exist in enormous variety with a huge gene pool, it is hard to imagine that Darwinian gradualism alone can ever lead to such reproductively isolated branches.

Of course, in part 1, I mentioned several more serious sources of dissatisfaction. A most important one is the existence of fossil gaps. A second important datum is the biological arrow of time. Biological evolution takes place in the direction of increasing complexity. In Darwinism, there is no room for fossil gaps and no room for the evolution in complexity.

So can we say then, with authority, that Darwin's theory is falsified, at least in part? Unfortunately, it is not so simple, because the framework of the theory is continually being adjusted—recall those Ptolemaic epicyles—though not without attendant problems.

First consider two small examples. Darwinism professes gradualism, slow rates of evolution that add up over long periods of time. However, when bacteria were found to develop resistance to insecticides, the Darwinians did not bat an eyelash before suggesting that the rate of mutation speeds up in those cases. But what about those species that display amazing stability, hardly changing at all through geological times? Does the variation rate trickle down to zero for them?

To explain why some species develop an adapted characteristic even before the environmental change requires it, the Darwinians invented the idea of preadaptation. The characteristic evolved by chance; then the environment also changed fortuitously and the characteristic got adapted. Nothing to it! But try to rationalize the appearance of a preadapted macroscopic trait that would require many hundreds of contributory variations by chance, all occurring without any environmental selection pressure. If you are not a believer, this leap is sure to boggle your mind.

Aside from these examples, four major additions to the original scope of Darwinism have been postulated to make the theory conform to crisis situations. The first is the idea of *geographical isolation*. When Eldredge and Gould (1972) suggested, on the basis of extensive fossil data, that there are two tempos of evolution, not only a gradual but also a fast tempo, the conservative response was to have recourse to the idea of geographical isolation (Mayr 1942). Because of geological changes, a small segment of a species may become isolated. Because of inbreeding, the isolated population may undergo rapid evolutionary changes. Later, if the two groups come together but are no longer able to interbreed, it would look like the sudden appearance of a new species. In this way, geographical isolation is used to explain away the fast tempo of evolution.

Prior to all the flap over fast-tempo evolution, geographical isolation was also used to explain away the inability of Darwin's theory to explain the concept of a species and speciation—change from one species to another.

The second addition is the idea of *neutral genes* and the related notion of *genetic drift*, which Darwinians use to explain why some species are maladapted (either underadapted or overadapted) to their environment. Originally proposed by Kimura and others (see Kimura 1983) to explain the fast tempo of evolution, neutral genes are "selectively neutral" and therefore can accumulate without being eliminated by natural selection (see also chapter 10). Kimura proposed that when enough neutral genes accumulate to make a new trait, natural selection may adapt the trait for its survival advantage, thus explaining macroevolution.

Darwinians were against the idea until they realized that accumulation of neutral genes could give rise to a genetic drift away from adaptation and thus explain maladaptation. Ernst Mayr (1980) sees evolution as beginning with small populations: "Favoring in every generation certain individuals owing to some properties they have, natural selection chooses automatically all their other genes, including many that are near neutral or even slightly deleterious. When such selection happens in very small gene pools, rather pronounced departures from the optimal genotype sometime survive owing to errors of sampling."

In this way, genetic drift has become a part and parcel of extended Darwinism, simultaneously giving it flexibility and making it more difficult to falsify. The biologist Colin Patterson comments:

Darwinian evolution, by natural selection, predicts that organisms are as they are because all their genes have been and are being subjected to selection, those that reduce the organism's success being eliminated, and those that enhance it being favored. This is scientific theory, for these predictions can be tested. "NonDarwinian" or random evolution [through genetic drift] predicts that some features of organisms are nonadaptive, having neutral or slightly negative survival value, and that the genes controlling such features are fluctuating randomly in the population, or have been fixed because at some time in the past the population went through a bottleneck, when it was greatly reduced. When these two theories are combined, as a general explanation of evolutionary change, that general theory is no longer testable. Take natural selection: no matter how many cases fail to yield to a natural selection analysis, the theory is not threatened, for it can always be said that these failures of selection theory are explained by genetic drift. And no matter how many supposed examples of genetic drift are shown to be due, after all, to natural selection, the neutral [gene] theory is not threatened, for it never pretended to explain all evolution. (Quoted in Hitching 1982, 63)

The third major extension was the idea of *genetic determinism*, which was first suggested by William Hamilton (1964) and popularized by Richard Dawkins (1976). Genetic determinism is the idea that everything in a biological system, be it a cell, a tissue, an organ, or the entire organism, is completely determined by the genes. This notion was invented to save Darwinism from a very important shortcoming: The chance variations take place at the gene (micro) level, but natural selection takes place at the phenotype (macro) level. In the absence of a straight, one-to-one connection between the genes and the macrolevel phenotype, the whole theory is dubious. With genetic determinism, when nature selects at the phenotype level, the selection automatically includes the relevant genetic variations.

That genetic determinism is a form of reductionism I have already mentioned, but here I want to emphasize that this reductionism is even more ambitious than the reductionism of physics and chemistry. If molecular biology were entirely identical with biochemistry, we would expect that the macro aspects of a cell or organ (for example) would be determined from the dynamics of the micro components of the cell or organ. But what kind of aspects? Only the law-like aspects, not the program-like aspects. Biological systems, in addition to law-like aspects, have a higher level of operation that I call program-like behavior in which they perform purposive functions that convey meaning.

Darwinists claim that, purposive functions not withstanding, all macroscopic traits and their functions are the cumulative results of slow evolution through chance and necessity. Yet materialist biologists believe that Darwin's theory, the chance-and-necessity mechanism of evolution, is also ultimately traceable to material laws. Isn't there a logical inconsistency here? If everything is ultimately traceable to law-like behavior, how can program-like behavior that belongs to a different logical category ever come about? Genetic determinism, in addition to implying physical (Newtonian) determinism, purports to save Darwinism from this logical inconsistency by asserting that the program-like aspects of the system at the somatic level are also determined by the system's genes, the genetic programs.

Normally programs require a programmer. Darwinism denies that. Instead, Darwinism is the epitome of materialist science: Genetic programs are the result of chance and necessity, especially natural selection. No programmer needed. In this way, biologists can maintain that all teleological statements of purpose in biology are ultimately teleonomical, purposive in appearance only. Program-like behavior at the macrolevel does not exist.

This is a neat little arrangement. Genetic determinism saves Darwinism from having to explain the micro-macro connection; Darwinism helps genetic determinism act as a truly versatile determinism reaching far beyond the scope of Newtonian determinism.

Genetic determinism, paired with Darwinism, becomes a huge black box in which all unexplainable problems of biology can be hidden. This box has, very appropriately, been called "Darwin's black box" by Michael Behe because it is really the combined result of Newtonian determinism and Darwinism.

Finally, the fourth addition deals with that pesky problem of the biological arrow of time. For quite a while, biologists ignored the problem by claiming that evolution does not necessarily take place in the direction of increasing complexity. But lately, the tune has changed because some biologists see in genetic determinism a window of opportunity for explaining the biological arrow of time. If the program-like behaviors of biological organisms come about via Darwinian adaptation, and if they all can be traced ultimately to genes, then why not postulate a gene (or genes) that programs the organism to evolve complexity itself?

These latest Darwinians have proposed a further modification of the original idea suggesting that evolution becomes "canalized" in a given direction because organisms adapt by developing *evolvability*—they learn to evolve (Kirschner and Gerhart 2005). A related concept is "evo-devo," that evolution and development go hand-in-hand, both being guided by the biology of genes (Carroll 2005). The canalization of the development of form is fairly well recognized, except that earlier work attributed this phenomenon to morphogenetic fields. Instead, suppose one asserts that the channels of development are due to the genes only; then doesn't it make sense to say that evolution uses these same channels through adaptation?

The hard truth is that, unless one is a true believer, these many epicycles of Darwinism do not satisfy. Even Ernest Mayr, an architect of the neo-Darwinian synthesis, has expressed doubts. The feud between the two most famous recent popularizers of biology, Stephen Jay Gould and Richard Dawkins, is well known. The "dialectic" schools of Richard Lewontin and the organismic school of Brian Goodwin continue to emphasize the lack of proper treatment of the environment in evolution theory. Biologists who have the guts to calculate probabilities can convince themselves most easily of the inadequacies of Darwinism in spite of the "cycles and epicycles," as some have done (see Shapiro 1987; Behe 1996). Others try to escape the self-imposed limits of biological theory imposed by Darwinism and genetic determinism through much broader holistic ideas of the Earth itself as a living system.

I think the reason most biologists don't make waves about Darwinism today is in part that they feel a reluctance to engage in politics, and in part that a credible alternative is lacking. Behe's book (1996) reads beautifully so long as he critiques Darwinism. But when he presents his own alternative, intelligent design, the book falls short. One viable idea—of "irreducible complexity"— does not a new paradigm make.

Quantum physics and primacy-of-consciousness thinking present, together, the first complete paradigmatic challenge to materialist biology and Darwinism. Our new approach, as you will see in chapters 11 and 12, can handle and integrate all the maverick ideas that biologists have put forth to *really* solve the anomalous data about evolution. Some of these maverick ideas I have already mentioned. In the next chapter I will talk about these ideas and others in a comprehensive manner. However, first we need to take a last look at the issue of whether Darwinism has truly been falsified.

THE MATTER OF FALSIFICATION AND THE CASE FOR A PARADIGM SHIFT

When I said that Darwin's theory is too general for comfort, what I meant is this: It is too slippery; it allows one to argue away its failure to explain data. That slipperiness, plus the fact that we have never seen nor can we expect to see any actual case of macroevolution in the laboratory, makes Darwin's theory very difficult to falsify.

The philosopher Karl Popper was so fed up with the situation in biology that he once proposed that all good scientific theories, by definition, should be falsifiable. That is, instead of using the conventional criterion of verifiability for the acceptance of a scientific theory, he said we should use the criterion of falsifiability. Darwin's theory is not falsifiable; therefore it is not an acceptable theory of science, according to Popper. His idea did not stick, but every Darwin skeptic would sympathize with his views!

So where does this leave us? Any scientific theory that passes the test of time has at least *some* data on its side. This is true of Darwin's theory; it was true of Ptolemy's theory. So some aspects of the scientific theory thus proven will stand.

But what happens when data pile up that require repeated apologies, excuses, and extensions of the theory—the now-famous "cycles and epicycles"? Well, we come to a criterion more practical than verifiability or falsifiability: namely, usefulness. We recognize that the theory no longer has any guiding power; there is no usefulness. That's when people begin ardently looking for a new, useful theory to replace it. A paradigm shift takes place.

Biologists have to come to terms with the fact Darwinism no longer guides research on biology—except for research by those few biologists who are looking for its replacement. Evolution research is popular with idea-impoverished philosophers, to be sure, and it feeds the ongoing political battle of science and religion, but hardly any graduate student comes to study biology with research in evolution in mind, unless he or she is a nonbeliever and intends to overthrow Darwinism.

Furthermore, one can rationalize away unfavorable data, but one cannot rationalize away the logical inconsistencies that also pile up for an aging paradigm. When that happens, the old paradigm has to let go.

It is time for Darwinism to go. I have already argued its logical inconsistencies. Let me recount them.

- Molecules and nonliving physical systems do not strive to survive. They do not have any self, any integrity to survive for. The concept of survivability is purposive, program-like. Survivability is not a logical criterion to use in a scientific theory that purports to be founded on materialism—primacy of matter and primacy of cause.

- The logical inconsistency above becomes very obvious when in one of the later extensions of Darwin's theory it is proposed that program-like behavior of biological beings is itself adapted behavior. It becomes a circularity of logic: Darwin's theory of adaptation requires survivability as the criterion, but survivability requires Darwinian adaptation.

- Some of the so-called evolutionarily adaptive behavior, self-awareness and the capacity for meaning processing, cannot even be processed by matter. So if you insist that this kind of behavior, including evolvability, arose from evolutionary adaptation, you are cutting off the very branch—material supremacy of being—on which you are sitting.

Biophysicists have already discovered the physical paradox-free basis of Darwin's theory. Instead of survival, we must use replicability or fecundity as the criterion of adaptation to an environment. Since not only living but also nonliving molecules have been found to replicate, replicability is a physical concept. So we define Darwinism—call it neo-neo-Darwinism—this way: Species that are most fecund or reproduce the best are the most fit to adapt to environmental changes. This redefinition makes Darwinism tautology free and logically sound. It also gives Darwinism the physical basis that biologists have long sought.

So now Darwin's theory is no longer slippery and subject to fanciful apologies and extensions. Does Darwinism defined this way agree with all experimental data? No. Even cursory observation shows that fecundity is a helpful criterion of evolutionary adaptation, but it is neither necessary nor sufficient for a species' ability to adapt and survive.

So Darwin's theory, taken in toto, is falsified.

But aspects of the theory are good and verified. Evolution happens; adaptation happens. Thus, creationists and proponents of intelligent design are wrong to dismiss Darwinism outright. Falsification may be a crucial test, as Karl Popper opined, but a good theory need not be completely rejected, emphasizing only its negatives.

The paradigm shift developed in this book is in line with all earlier paradigm shifts in science. The old biology—Darwinism and molecular biology—is retained as a limiting case, a case in which the theory applies whenever certain limiting conditions are met.

Having said that, we must also recognize that to make evolution research useful once again in mainstream biology we must join the effort toward a paradigm shift. The upcoming paradigm shift is more like the shift from the old Greek science of Ptolemy and Aristotle to the modern views of Copernicus, Galileo, and Newton. It is less like the relatively benign transfer of leadership from Newtonian physics to Einsteinian relativity or even from Newtonian physics to quantum physics.

Naturally, the stakes are high, and politics cannot be avoided. The organismic and holist opposition in biology acts like a faithful, civilized opposition party. The organismic biologists try to modify biology toward holism while keeping the idea of materialism—the primacy of material—unchallenged. In this way they fail also in their search for a useful paradigm for a new biology. Holism does not go far enough; a more radical idea is needed. We do have to go back and revive some of the historical debate that took place between the causalists and teleologists, between the supporters of adaptive conditioning and those of biological creativity, between upholders of upward causation and of downward causation. We do have to develop a biology within the primacy of consciousness, politics not withstanding.

In some sense, this paradigm shift and the associated muddy politics are happening already as part of the movement of our consciousness. Witness the growing intensity with which the idea of intelligent design is being opposed by materialists everywhere, not just by the biologists. In a *Doonesbury* cartoon, one of the characters says, "Drat! These pesky scientific facts won't line up behind my beliefs." Darwinists often use this kind of line when depicting creationists and intelligent designers. But if you are not a believer, you can see, upon reading this chapter, that this line also fits Darwinism pretty well. Both camps lack credibility about explaining *all* the "pesky scientific facts."

The materialist philosopher Daniel Dennett (1995) wrote a book entitled *Darwin's Dangerous Idea*. Darwin's idea *was* dangerous, although not in the way that Dennett meant it. It was dangerous because it enabled materialists to hold on to and perpetuate a false metaphysics for far too long. For the five decades since the discovery of DNA, the field of biology, the science of life, has been guided chiefly by a philosophy that applies (and even then approximately) only to the nonliving. Isn't it time for materialism in biology to go?

10

POST-DARWINIAN IDEAS
and the JOURNEY to the
NEW PARADIGM

O ne has to admit that until quantum physics came along, the idea of biological creativity could not have been made into a solid theory. In that respect, the pragmatism displayed by the biological leadership in advocating Darwinism is not entirely unjustified. However, this seductive pragmatism has lulled biologists and even the public into thinking of evolution and Darwinism as synonymous. This is no way of doing science, to lift an at-best partially verified theory to the status of gospel. Fortunately, scientific skepticism gradually began to break the spell of pragmatic Darwinism and open minds to other ways of viewing evolution.

In this chapter I trace a brief history of biologists' thinking on ideas that are precursors to the idea of biological creativity à la quantum physics. I will also discuss the debate around these ideas and how these debates can be settled by using the idea of biological creativity, which will provide the context for the next chapter.

I will also try to convey to you a sense of how these debates have always been settled by the mainstream biological community— not through the normal channels of scientific data or breakthrough in theory building but through political realism. Darwinism has always been the compromise position: Since no other theory has been compelling, why upset the apple cart? There is always the threat of politics in the background, both within the discipline and in the ongoing debate with the believers of biblical Genesis, and this threat seems to have prevented a more honest appraisal of the situation.

From within the political melee, some promising noises of post-Darwinist thinking can be heard; they are the focus of this chapter. One streak of anti-Darwinian thinking is well publicized: the new Larmarckism. Some of the history of this idea I have reserved for a later chapter on the new-paradigm thinking on Lamarckism (chapter 20).

A second important streak in post-Darwinian thinking focuses on gradualism. Biologists have always known gradualism is not adequate to explain the appearance of macroscopic traits and maybe even of speciation. The idea of discontinuous biological creativity took shape fairly early. The idea of two tempos of evolution also began to appear. Anti-Darwinian ideas such as "hopeful monsters" and "neutral genes" were floated, and the latter was even co-opted by the Darwinists to their advantage.

Several other themes have also begun to emerge. One is an acknowledgment of the importance in evolution of catastrophes that led to large-scale extinction of species. Nobody can deny those extinctions, however contrary to Darwinism they may be. Many biologists have also noted how a flurry of evolution followed these extinction events. And we must not forget the organismic perspective when considering creativity in evolution. Are there patterns or laws of form that guide the final shaping of creativity in evolution? Finally, responding to one of the most glaring cases of the impotence of Darwinism, the evolution of eukaryotes from the early prokaryotes, the biologist Lynn Margulis (Margulis and Sagan 1986) suggested a cooperative mechanism in contrast to Darwinian competition. In the same vein, James Lovelock (1982) envisioned large-scale cooperation in the form of a planetary Gaia-consciousness that has important consequences for evolution. In the next section, we embark on a tour of some of these post-Darwinian ideas.

HOPEFUL MONSTERS

The biologist Richard Goldschmidt was a refugee from Hitler's Germany who settled at the University of California, Berkeley, to continue his work. His most famous idea, that of "hopeful monsters," was proposed in 1940 (see Goldschmidt 1952), a time when the neo-Darwinian synthesis was being drafted. Goldschmidt was one of the few biologists who, while accepting the plausibility of Darwinism in explaining microevolution, had grave doubts about the ability of the theory to explain macroevolution and hence doubts abut the viability of neo-Darwinism. In an oft-quoted challenge to the neo-Darwinists, he said:

> I may challenge the adherents of the strictly Darwinian view . . . to try to explain the evolution of the following features by accumulation and selection of small mutants: hair in mammals, feathers in birds, segmentation of arthropods and vertebrates, the transformation of the gill arches in phylogeny including the aortic arches, muscles, nerves, etc.; further, teeth, shells of mollusks, ectoskeletons, compound eyes, blood circulation, alternations of generations, statocysts, ambulacral systems of echinoderms, pedicellaria of the same, cnidocysts, poison apparatus of snakes, whalebone and finally chemical differences haemoglobin versus haemocyanin, etc. . . . Corresponding examples from plants could be given.

Never mind that as a nonbiologist you may be unfamiliar with some of the biological terms in the quote above (this author himself had to look up several of them). Just appreciate that even today, Darwinists hand-wave their way to explanations of such features of macroevolution, a point that Michael Behe (1996) makes in detail about some similar features.

Being a courageous man, Goldschmidt proposed that such macroevolutionary changes happen suddenly, via "monstrous" mutations. You may have seen phenotype-level exhibitions of such mutations in fairgrounds, in a display of a two-headed sheep or the like. Of course, most often such mutations would not survive, but maybe, on rare occasions, Goldschmidt argued, they do survive, and that survival leads to macroevolution and new species. To such surviving monstrosities of simultaneous genetic variation Goldschmidt gave the evocative name "hopeful monsters."

The idea is preposterous, Darwinists have said, then and now, and Goldschmidt hardly gets a footnote, if that, in today's biology textbooks. But he has become quite a famous name in the small anti-Darwinist camp. For example, in 1975, the biologist Guy Bush of the University of Texas endorsed Goldschmidt's work with these lines: "Goldschmidt's 'hopeful monster,' a mutation that, in a single genetic step, simultaneously permits the occupation of a new niche and the development of reproductive isolation, no longer seems entirely unacceptable."

The idea is preposterous when viewed from a classical-physics prejudice of continuity. But look at the idea from a quantum view and hopeful monsters will become a very hopeful idea. The quantum view takes you to the "preposterous" idea that discontinuous quantum leaps enable electrons to travel to a final destination without going through the intervening space. Like the movement of electrons, gene mutations are quantum events, too; they are, after all, possibilities. What prevents these possibilities from accumulating without Darwinian selective scrutiny by nature? Only a classical freeze-up of the imagination. If you can defreeze your mind's eye, the idea of hopeful monsters can become plausible. Then you can appreciate the value of Goldschmidt's far-sighted intuition, an insight way ahead of its time.

Neutral Genes

In the late 1960s and early 70s another idea appeared in the biological literature that can be seen as a precursor to quantum thinking (Kimura 1983): the idea of "neutral" gene mutations. It is important to remember that many gene mutations are neither beneficial nor harmful. Now recall that natural selection eliminates harmful mutations and selects beneficial mutations. But if a mutation is selectively neutral—neither beneficial nor harmful—natural selection does not act on it. Can such variations in the gene pool make a difference in how a species evolves?

Consider some probabilities. Such a variation has a one-in-two chance of surviving to the next generation; with two parents contributing genes, that much is obvious. Calculations show that there is only one chance in a thousand that such a variation will survive to the thousandth generation.

But probability theory has enough openness to suggest that by chance alone, some small number of such neutral variations can become established in a species population despite the odds. Probability theory also allows us to show that for large populations, the presence of neutral genes is unlikely; large populations tend to arrive at an equilibrium in genotype frequencies. Biologists call this the *Hardy-Weinberg equilibrium.*

But the probability estimates leave the situation quite open for small populations, in which, consequently, a genetic drift away from the Hardy-Weinberg equilibrium may occur. Darwinists have used the idea of genetic drift as a loophole to get out of many embarrassing situations, as I discussed in the last chapter.

However, probability estimates also admit of yet another

possibility, the scenario that makes the neutral gene hypothesis a precursor to quantum thinking. It seems possible that neutral variations can lie dormant in various individuals of a population. It also seems possible that every once in a while such neutral variations may converge in the sexual reproduction of a new individual in whom they are activated, in the sense that they can produce an adaptive trait. In other words, all of a sudden, genes that were previously unexpressed can be expressed, thus explaining the sudden appearance of a hopeful monster with a new trait.

In this reconstruction of events, the previously discussed evolution of the giraffe's neck can be interpreted as follows. Suppose that, over a period of thousands of generations, a series of neutral mutations accumulate, mutations that can potentially produce stronger bronchial arches, bigger muscles, and a bigger heart. But in the absence of a long neck such mutations have no use and mean little harm, so they may become fixed in a population, small probability and all. Now suppose that these mutations all converge in a few new offspring in which there is also the crucial new mutation that is responsible for the long neck. All of a sudden, the support system for the neck is expressed synchronously with the long neck. This interpretation gives much-needed credibility to the evolutionary explanation for the neck of a giraffe.

But is it really any more credible than the normal Darwinist scenario? Is it really possible for genes for a bigger heart or bigger muscles to go selectively unnoticed?

Look at the idea with quantum thinking: Gene mutations are mere possibilities, not actualities, until the event of collapse. Nature, in the unconscious state assumed by Darwinism, cannot select from possibilities. In this way, possible gene mutations can accumulate unselected until their time comes, as in the above scenario.

In the creativity literature, this part of the creative process is called *unconscious processing*, something that is unallowed in classical thinking but completely kosher in quantum thinking. This notion of unconscious processing is discussed further in chapter 12.

Catastrophes and Extinction

Darwinism makes an important implicit assumption: For evolution to be gradual and continuous, the environment whose selection drives the evolution must also change gradually and continuously. This hypothesis goes by the unwieldy name of *uniformitarianism*.

Unfortunately, geological data clearly indicate that the environment has occasionally been anything but uniform. Earth has undergone quite a few major upheavals. According to natural historian Norman Newell, these major upheavals took place at the beginning and end of the Cambrian period, and at the ends of the Devonian, Permian, Triassic, and Cretaceous. In fact, they help to set the boundaries of the geological time scale.

If the environment on Earth goes through such catastrophic events, how can we expect the evolution of organisms to be continuous or claim that extinction of a species is due to its unfitness or maladaptedness? Indeed, these catastrophic events have been identified as epochs of mass extinction followed by enormous explosions of new life.

In the massive extinction that took place at the end of the Permian period 225 million years ago, some ninety-six percent of the existing species became extinct. Did the other four percent survive because they were the fittest? "There are few defences against a catastrophe of such magnitude," said Gould, commenting on this catastrophe, "and survivors may simply be among the lucky four per cent . . . our current panoply of major designs may not represent a set of best adaptations, but fortunate survivors." And Newell said this:

> . . . the fossil record of past life is not a simple chronology of uniformly evolving organisms. The record is prevailingly one of erratic, often abrupt changes of environment, varying rates of evolution, extermination and repopulation. . . . Mass extinction, rapid migration and consequent disruption of biological

equilibrium on both a local and worldwide scale have accompanied continual environmental changes." (Quoted in Hitching 1982, 136)

This kind of realistic reconstruction of the fossil record led to many non-Darwinian alternatives to evolutionary theory. Hitching (1982) constructs a generic version of these theories for the evolution of land animals:

In the wake of a disaster (probably global) a large number of amphibious creatures were thrown far up on shore and became stranded. Many died of starvation and injury, but many also survived, and among the mothers the multiple effect of changed diet, stress, prolonged exposure to a new climate, and acquired immunity to virus diseases led to intense genetic pressure to their unborn young. Chromosomal changes led to hopeful monsters in their thousands emerging from new-laid eggs. The vast majority were stillborn, or impotent, or failed to make an impact because the chromosomal change was "bred out." But just occasionally, through harem-type breeding in isolated populations, the chromosome change was perpetuated. After a few generations, several varieties of "monsters" became viable— new species which then proliferated over a largely unpopulated globe. (143)

You can see that there is a certain credibility in this kind of scenario, particularly when you realize that the only response Darwinian gradualism can make to the cacophony of such catastrophes is "uh, the survival of the luckiest." By adding the quantum notions presented thus far, the credibility manifestly increases. It is neither the fittest nor the luckiest that make the grade when a massive catastrophe takes place; it is the most creative.

The truth is, certain groups of organisms suddenly come into play, remain homeostatic though millions of years, and then suddenly die out. In other words, evolution has both Darwinian and non-Darwinian epochs. This idea has been further codified in the previously mentioned idea of slow and fast tempos of evolution.

Two Tempos of Evolution: Punctuated Equilibrium

The idea of two tempos of evolution is sometimes called *punctuated equilibrium*. The Darwinian, slow-tempo, continuous evolution is the equilibrium, something akin to the continuous prose of a narrative; the "quantum," fast-tempo evolution is like punctuation marks in the narrative.

Incidentally, biologists themselves have used the word *quantum* as an adjective for the fast-tempo evolution (Grant 1977). Unfortunately, perhaps because of political pressure, even the original proponents of the idea, Stephen Jay Gould and Niles Eldredge, have compromised on the quantum-eye view of the situation and retreated to a preference for a Newtonian-eye view. For example, in Gould's early work (Gould 1980), one finds the word *discontinuity* and the idea that during the two tempos, even the basic processes of variation and selection "work in different ways." However, in the later writings of both Gould and Eldredge, one is disappointed to find that these authors have resigned themselves to accepting Ernst Mayr's explanation of the punctuation marks of the fast tempo as being due simply to geographical isolation.

If you are not a believer of Darwinism and have no political axe to grind, you can see that the fast tempo can be very fast indeed: It can be symptomatic of instantaneous quantum leaps. In such leaps Goldschmidt's hopeful monsters come alive through the accumulation of quantum possibilities of Kimura's neutral gene variations (see chapter 11).

ORGANISMIC BIOLOGY; OR, WHAT SHAPES FORM IN BIOLOGICAL CREATIVITY?

There are one hundred thousand species of butterflies and moths. The biologist Frederick Nijhout, who has studied them extensively, has been able to resolve the resultant one hundred thousand patterns of form to about six or so basic forms. What does this imply? Nijhout explains:

> Few things in nature match the beauty and the variety of patterns on the wings of butterflies and moths. This order of insects—the lepidoptera—consists of some 100,000 species, and virtually every one of them can be distinguished from the rest solely by the color pattern of its wings. The phenomenon is even more remarkable when one examines how patterns are formed, as I have done over a period of years, and finds that the answer is essentially rather simple. . . . The development of color patterns can be achieved, at least in principle, by the same kinds of processes that guide the development of morphological features, because all development is ultimately the outcome of progressive changes in the expression of genes. (Quoted in Augros and Stanciu 1988, 192–93.)

In chapter 4, I pointed out that "organismic" biologists contend that there must be processes that guide the development of macroscopic forms and functions—morphological features, patterns of color, functions of organs, and the like—at the phenotype level, starting with the expressions of the genes, that is, the proteins. They further contend that such laws or processes must play a role in evolution. Organismic ideas of evolution can be traced back to such luminaries as Baron Cuvier and Hans Driesch, biologists working around the turn of the nineteenth and twentieth centuries, respectively.

Biologists thinking in this vein have found several weak spots in traditional notions of form. In Darwinism, form is due to genes, and therefore, ultimately, form is the product of mere chance. This approach, said Driesch at the turn of the twentieth century, "explains how by throwing stones one could build houses of a typical style." Darwinism concentrates wholly on genes. However, because (even according to Darwinism) natural selection acts on the phenotype, the expression of the genes, and not on the genetic level itself, the contention of the organismic biologists makes sense. In Darwinism, the dominant assumption is that the genes have the entire recipe to build form. In organismic theory, the genes have the recipe to build not form, but proteins; the road from proteins to cells, from cells to organs, requires further guides.

One problem that illustrates the importance of considering form in evolution is the phenomenon of homology, which is often cited as a major proof of evolution itself. The term refers to the fact that organisms of very different types possess homologous organs; that is, the same pattern of form seems to have evolved to perform many different functions. For an example, consider the tetrapod form in vertebrates: The flippers of a whale, the legs of a camel, and the wings of a bird are all built in the same basic form, although they serve quite different functions. If evolution were the fittest adaptation of gradual accumulation of form generated from chance variations, one would expect that the forms used to swim, to run, and to fly would differ greatly from each other. Compounding the puzzle, indeed, the gene complexes responsible for the respective limbs are quite different and yet, somehow, these different genetic systems have managed to produce the same basic forms. This homology is convincing evidence that there must be laws of form or processes that guide form.

Initially these guiding principles, or morphogenetic fields, that help generate form were considered to be local. Conrad Waddington thought they operated through the canalization of form building. More recently, Darwinists have tried to introduce the notion that canals for form making have come about through evolutionary adaptation. In 1981, Rupert Sheldrake added the concept of nonlocal morphogenetic fields, which can only be non-physical. I have already discussed, in chapter 4, how this concept revivifies the idea of the vital body, an ancient "bottle" now serv-ing the new "wine" of the blueprints of forms and functions.

From an entirely different context, the physicist Ilya Prigogine and later chaos theorists have looked at biological evolution as evolution of order within chaos. I previously mentioned these ideas in connection with the origin of life (chapter 6). There the main problem is how to produce program-like behavior of a living cell from the law-like behavior of atoms and molecules, for which order within chaos does not seem to be a productive idea.

However, once program-like behavior is available as the basic building block of the living cell, the basic living form that carries a vital program, and once we understand the process of evolution as the creative evolution of more and more sophisticated vital pro-grams, then seeing form building as creating order within chaos has considerable merit (see chapter 11).

GAIA THEORY AND SYMBIOSIS

How does one figure out whether there might be life on a planet without having to detect it directly? When Mars scientist James Lovelock had to deal with this problem in connection with a Mars expedition, the answer he came up with was that to sustain life a system has to be in thermodynamic *dis*equilibrium. A state of thermal equilibrium would mean that the entropy or disorder of the system has already been maximized, which means in turn that all creation of order has ceased. A little thought gave Lovelock the further clue that the planet's atmosphere should deviate from chemical equilibrium. Indeed, when Lovelock looked at data, he found that Earth's atmosphere has a very odd mixture of gases when compared to Mars or Venus, a mixture in "exuberant disequilibrium" compared to relative equilibrium.

When he took his idea to astrophysicist Carl Sagan, he found out another peculiarity of the life-bearing Earth: that Earth has maintained stable temperatures through billions of years, whereas theoretical calculations show that the sun has heated up considerably during that period. These two pieces of peculiar data eventually gave rise to an amazing insight in Lovelock's mind: The Earth must be a living organism able to regulate its temperature and atmospheric chemistry. Thus was born the now-famous "Gaia hypothesis" (Lovelock 1982).

The biologist Lynn Margulis (1993) enriched the Gaia hypothesis by calling attention to the important symbiosis between the environment and the biota of the planet. A bee has a symbiotic relationship with the flowering tree for which it spreads pollen while deriving its food. Margulis also originated the idea that such symbiotic relations on the part of microorganisms of bacteria with their hosts play an important role in evolution. Despite being a non-Darwinian idea, it has gained some acceptance from the establishment through the sheer weight of data Margulis can present. Margulis theorized that the transition from the prokaryotic cell to the nucleated eukaryotic cell occurred through such symbiotic hosting.

The Gaia hypothesis has, over the years, gained many supporters, though mostly outside biology. Biology insiders remain skeptical, partly because the idea is a little vague, but mostly because it smacks of holism, in the form of an emergent planetary consciousness. Certainly it proposes a bigger role for life (and consciousness, by inference) and on a much larger scale than that claimed by the materialists and reductionists, who refuse to see any significant difference between life and nonlife.

Because it proposes a role for consciousness, the Gaia hypothesis in biology has helped support the idea of science within consciousness in general. In chapter 12, I will show that when we

reformulate a theory of evolution within the primacy of consciousness, the ideas of Lovelock and Margulis fit in beautifully.

DOES CONSCIOUSNESS CREATE
BIOLOGICAL ORDER?

Finally, I must mention a very important book by the philosopher
Robert Augros and the physicist George Stanciu (1988), who were
probably the first scientific researchers to suggest that consciousness plays the crucial role in the creation of life and life's evolution. I will let them speak for themselves:

> What cause is responsible for the origin of the genetic code and
> directs it to produce animal and plant species? It cannot be
> matter because of itself matter has no inclination to these
> forms, any more than it has to the form Poseidon, or to the
> form of a microchip or any other artifact. There must be a cause
> apart from matter that is able to shape and direct matter. Is
> there anything in our experience like this? Yes, there is: our own
> minds ["consciousness" in the language of the present book].
> The statue's form originates in the mind of the artist, who then
> subsequently shapes matter, in the appropriate way. The artist's
> mind is the ultimate cause of that form existing in matter, even
> if he or she invents a machine to manufacture the statues. For
> the same reasons there must be a mind that directs and shapes
> matter into organic forms. Even if it does so by creating chem-
> ical mechanisms to carry out the task with autonomy, this artist
> will be the ultimate cause of those forms existing in matter. The
> artist is God and nature is God's handiwork. (191)

In my opinion, however, Augros and Stanciu did not quite
achieve their goal in their book primarily because quantum physics and downward causation by consciousness were not available to

them as the conceptual framework around which to build their theory. With these concepts available, the astrophysicist Arne Wyller (1999) was the first to invoke consciousness as a solution to all the problems that Darwin's theory does not properly address. However, Wyller's work lacked any significant details. The present book fills in the details and completes the vision of these three authors. The next chapter is the denouement: We are finally ready to introduce the concept of creative evolution in some detail.

11

CREATIVE EVOLUTION

In part 1, I drew the outlines of a new approach to biological evolution that I call creative evolution. The essential idea is that new species most often arise through creative acts of consciousness. This biological creativity takes place in quantum leaps, instantaneously. I submit this creativity as the elusive cause of fast-tempo evolution that biologists have been seeking for decades. In this way, via the fossil gaps, evolution proves intelligent design and a creative designer.

In this chapter and the next I show that creative evolution integrates the concerns of practically all the viewpoints mentioned thus far. Proponent of intelligent design? The designer is quantum consciousness/God. Granted, the designer is not exactly the God of Christian belief, but is the difference truly significant? Punctuation theorist? It takes one giant, discontinuous leap—creativity— to manifest all the right genetic possibilities for the raw material for making a new form. What a punctuation mark! Advocate of hopeful monsters or neutral genes? Amazingly, quantum creativity incorporates both ideas in one fell swoop. Organismic biologist? Making the novel form requires invoking morphogenetic fields and laws of form. Catastrophe thinker? Catastrophes are part of creativity; some destructuring is needed before creation to open ground for the play of the new. Not least, neo-Darwinist? In the new theory, Darwin's slow mechanism is found to be the situational, conditioned (correspondence) limit of God's creative downward causation (see chapter 12). Moreover, the creator God is an objective God that passes the stringent test of a scientific theory.

MACROEVOLUTION AND QUANTUM LEAPING

The biggest problem in explaining biological macroevolution is this: A macroevolutionary change requires so many micro changes at the genetic level, so many mutations! This difficulty has motivated the notions of punctuated evolution, hopeful monsters, neutral genes. Let's see how those theories can be integrated through creative evolution.

First, let's visit a favorite example, the eye. The evolution of a retina from a primitive light-sensitive spot requires many coordinated changes, literally thousands upon thousands of new genes, all of whose effects must be coordinated. That's just the eye itself; a retina backed up by brain processing is yet another matter. However, each gene variation, according to Darwinism, must be selected individually. The likelihood of single variations being beneficial is minuscule; in fact, they're more likely to be harmful. So there's a good chance individual selection would remove most gene variations. Beyond genes, there are more issues. It has become clear that producing the whole organ requires additional, somatic-level programming. Accumulating so many beneficial gene variations and so much programming at the somatic level would take far longer than geologic time! A quantum leap is the only solution that makes sense.

To see how quantum creativity works to produce macroevolutionary jumps, let's first explore the paradox known as Maxwell's demon, named for its creator, the mid-nineteenth-century physicist James Clerk Maxwell. Imagine that a demon wants to sort out slow molecules from fast molecules in an ensemble of molecules that initially has a random mix of speeds, a situation physicists call *thermal equilibrium*. The demon puts a partition with a trapdoor in the middle of the box of molecules and sits at the trapdoor (fig. 9). Whenever the demon sees a fast molecule moving from left to right at the trapdoor, it lets it go through. And whenever a slow molecule is ready to pass through the trapdoor from right to left, again the demon lets it through. What began as a random mix

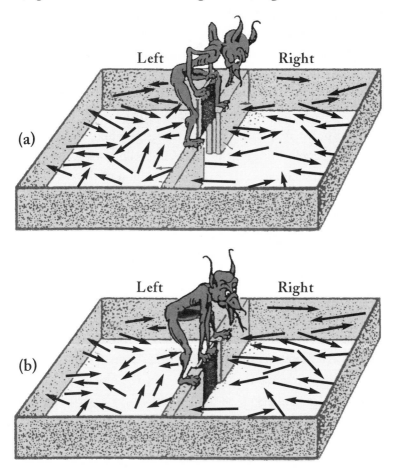

Figure 9. Maxwell's demon. (a) The molecules of both chambers are disorderly, both slow (short arrows) and fast (long arrows) molecules. (b) The demon has produced order in both chambers by his trapdoor trick: the left chamber now has only slow molecules; the right chamber has only fast molecules.

is now two regions with different characteristics—more fast molecules on the right, more slow molecules on the left. So, by this trapdoor trick, the demon has created order from disorder.

Now recall the entropy law: In natural processes, entropy, or disorder, always increases, or at best stays the same. Maxwell thought up this little scenario to demonstrate a possible violation of the entropy law—a paradox. Is the paradox real? No. Another physicist, Leon Brillouin, resolved the paradox. He pointed out that every time the demon looks to see whether a molecule is the right one to let through the trap door, it has to spend energy to get that information. The price of using energy is entropy, so entropy increases overall. The creation of order—say, the creation of life forms of increasing complexity—costs entropy.

This point is favored by creationists, who complain that, because the evolution of life is an uphill battle against the entropy law, evolution is impossible, being in violation of the entropy law. The evolutionist retorts, No, my friend, there is no violation. First, the Earth is an open system; it receives energy from the sun to drive the uphill battle against entropy. Second, although organisms emerge, the entropy of the environment around them increases, more than offsetting the decrease of entropy due to the evolution of life.

It sounds like a good answer, but there is a problem. Although the argument holds for maintaining life (Schrödinger 1944), there's a difference between "maintaining" and "evolving." For maintenance mode, we know how it works: how the biota on Earth take in solar energy and also how the entropy of the environment goes up in the process. For evolution, we do not know the "how" of either process. In other words, the evolutionists' argument that the entropy law is not violated by evolution is plausible, but no more than that. It does not completely satisfy.

Can creative evolution provide a better answer to the entropy question? I am proposing an evolutionary mechanism based on quantum creativity—creation by a quantum consciousness. In the mechanistic model of creativity that follows Newtonian physics, a creative "insight" is the result of conscious processing, via trial and error. In fact, can we not see Maxwell's demon as a conscious processor of this ilk? Thus, in conscious processing, the same concern about entropy applies. In quantum creativity, however, the situation is saved because the creative insight is preceded by unconscious processing—processing without awareness—that requires no dissipative energy and does not increase entropy.

In quantum thinking, the gene mutations are quantum possibilities (Elsasser 1981). In the quantum world, with its ever-shifting landscape of potentia, these possibilities, unless collapsed, will interact, expand, and accumulate. Biologists who think in a traditional vein may want to assume at this point that the quantum gene variations collapse by themselves as they arise without any help (downward causation) from consciousness. As discussed earlier (chapter 2), though, consciousness and its power of downward causation are required if the quantum collapse event is to be free of logical paradoxes.

Any gene variation present in possibility that is not expressed in creating a macroscopic trait remains uncollapsed. Consciousness does not collapse the unexpressed gene variations—all consisting of quantum possibilities—until a whole gestalt of them, when expressed, would make a new form. As we do in our own creative process, consciousness waits for the right creative moment.

Does this sound familiar? It's Kimura's elegant idea of neutral genes translated into the language of the quantum world, where the idea can become far more complete. In the quantum view, unexpressed gene variations are neutral with respect to selection because they are not manifest. By the same token, Darwinists are now denied the use of the neutral gene idea as an escape route: They can no longer invoke genetic drift whenever a species is imperfectly adapted to the environment (see chapter 10).

Does this mean genetic drifts are not allowed in creative evolution? Not necessarily. Human creatives are known to make false leaps, having misread the gestalt of possibilities. Why shouldn't biological creativity also evince occasional false leaps that manifest a bunch of genes that don't quite add up to a new trait? Such false leaps are responsible for what we observe as genetic drift.

Like Kimura, Goldschmidt also had a sneak preview of quantum creativity in evolution when he conceived the hopeful monster. It is extremely hard to credit the idea that a hopeful monster could be produced by chance. But if the radically new form — the hopeful monster — is produced by unconscious processing and a quantum leap of insight. . . ? In those terms, the idea becomes credible.

We also need to remember that the radically new, manifest form is not in fact arrived at by itself: The corresponding vital blueprint is also available to the unconscious processor that is quantum consciousness/God. That blueprint offers a rough guide-line of what needs to be sought through unconscious processing.

Let's get back to the central question, the leap itself. When does consciousness choose? Well, before any choice can be made, consciousness needs microlevel possibilities to be amplified into macrolevel possibilities. Therefore, collapse does not take place at the micro genetic level. An amplification of the micro genotype to the macro phenotype first takes place in possibility. I think that this amplification involves the chaos dynamics that I mentioned in chapter 6; this idea is explored in a later section.

When there is a match between the possibilities for macro-physical form and the morphogenetic blueprint of form, a match that Rupert Sheldrake (1981) calls *morphic resonance*, collapse of the possibility waves precipitates, a quantum leap takes place all at once, and consciousness has succeeded in making a physi-cal representation (the physical trait or organ, the form) of the morphogenetic blueprint and, along with it, a new species or even higher taxon. *There are no fossil records for the intermediate stages, because there are no manifest intermediate stages!* It is as simple as that.

In trying to escape just such conclusions, a Darwinist tends
to act like a magician trying to create an illusion, in this case,
the illusion of gradualism. Look at the biologist Gavin de Beer's
depiction of the eye in his 1964 *Atlas of Evolution* (fig. 10). The
image on the left shows a one-celled organism with a light-
sensitive spot and a primitive lens. The middle image is a section
through the eye of a jellyfish, showing the lens and light-sensitive
cells forming a cup-shaped retina. At the right, the eye of a tad-
pole is shown with the lens and light-sensitive cells of the retina,
but now the retina is formed from the lining of the brain cav-
ity. Gradual evolution? Hardly. The evolution of a retina from a
primitive light-sensitive spot requires many coordinated changes,
and it takes a quantum leap to achieve them. From this primitive
retina, the evolution of a retina backed up by brain processing is
a still bigger quantum leap.

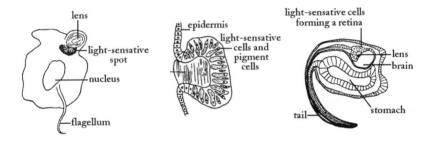

Figure 10. Three stages of the evolution of the eye, from Sir Gavin
de Beer's *Atlas of Evolution* (1964). The stages don't prove
gradualism as Sir Gavin purported (see text).

CREATIVE EVOLUTION AND
THE INTERMEDIATES

How then should we interpret the few cases that Darwinists hail
as clear evidence for intermediates in a gradual evolution? Let's

examine one famous case of one so-called intermediate—the archaeopteryx.

When the archaeopteryx was discovered, it was immediately hailed by the Darwinists as an intermediate between reptiles and birds. This species seems to have both fully developed feathers and clear reptilian features, such as a long, tooth-equipped jaw and trailing tail. Unfortunately, as biologists themselves have noted, the creature lacks the strong sternum to anchor the powerful muscles that operate the wings in birds. Therefore some researchers have concluded that archaeopteryx did not fly.

What are the feathers for, then? In Darwinism, where every feature has to be an adapted feature with survival value, these feathers are an embarrassment! Recall the problem of preadaptation that haunts Darwinists: How can an adapted feature appear before it is used? They suggest that the feathers of archaeopteryx were not for flying but for conserving heat, overlooking the fact that the feathers were flight feathers, not insulating down feathers.

Creative evolution has a very simple answer for all cases of preadaptation. The preadapted form is indeed arrived at via a quantum leap, but further adjustments are needed to make the creative new trait useful for the purpose for which it was designed. This is the well-established manifestation stage of creativity. In other words, an archaeopteryx has feathers suitable for flying; therefore, it is unambiguously a bird. After all, researchers do agree that it should have been able to fly as high as the branch of a tree to save itself from a predator. To be sure, the archaeopteryx still has to develop a strong sternum, muscles supporting flight, and perhaps even suitable brain power before it can fly properly like a modern bird, but why deny that it is a bird? By assigning an implausible, intermediate use to a highly complex, prized feature, Darwinists do an injustice to the powers of nature.

I hope you see from this example that the fossil gaps can in fact be viewed not as something that has to be explained away but as valuable data. They are evidence for real discontinuity. These gaps will never be filled through future discoveries because they are evidence of biological creativity, of quantum leaps in evolution. As such they provide us with some of the most spectacular evidence for downward causation and God's intelligent design.

One question still needs to be addressed. In our own lives, creativity consists of the individual human creative taking a quantum leap to quantum consciousness/God, making the creative quantum collapse possible (Goswami 1999). Clearly, the individual consciousness has a role to play. What is the role of the organism in biological creativity? We will return to this question later in this chapter.

LAWS OF FORM:
CHAOS DYNAMICS

I noted before that the laws of macrophysical form most likely involve chaos dynamics. Let's elaborate.

First, consider how we approach the idea of macrolevel form in creative evolution (fig. 11). On one hand, we have an organism with certain macroscopic quantum possibilities of form arising from the dynamics of matter (which in turn originate from the quantum possibility inherent in microscopic genetic variations). On the other hand, we have the vital blueprint of form, which consciousness also processes to represent in matter. Creative insight in evolution happens when the creating consciousness finds a match between these two types of form.

Notice the large and important step implied there: Microquantum possibilities at the level of genes are amplified to macroquantum possibilities at the level of the organism. To understand this step, we need to look at the dynamical laws of form.

Biologists, even those we call organismic biologists, think of the laws of form as deterministic: Given the ingredients, the

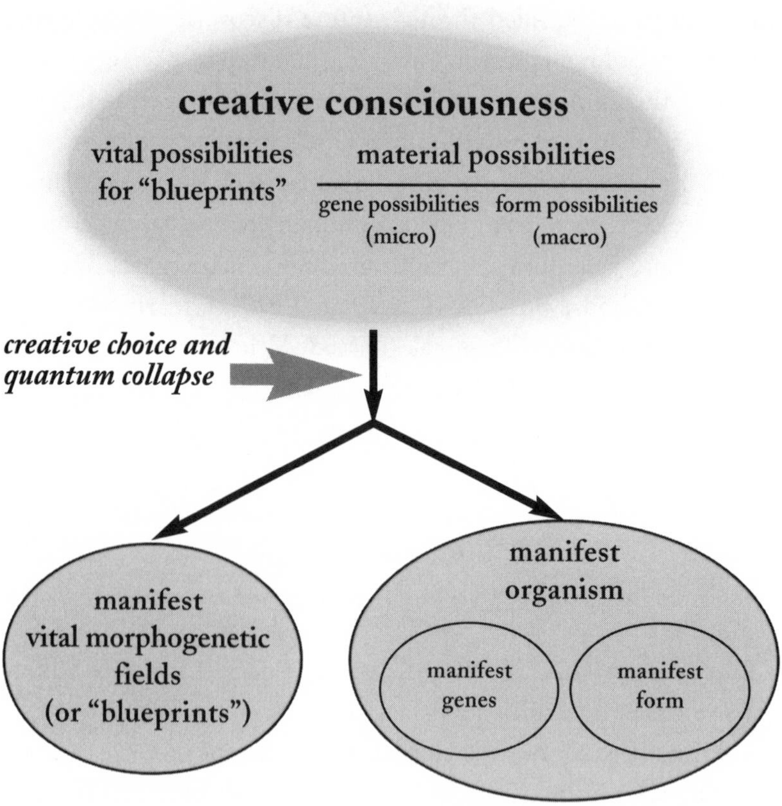

creative consciousness

vital possibilities
for "blueprints"

material possibilities

gene possibilities   form possibilities
(micro)   (macro)

creative choice and
quantum collapse

manifest
organism

manifest
vital morphogenetic
fields
(or "blueprints")

manifest
genes

manifest
form

Figure 11. The relationships between consciousness and possibilities
of form in the theory of creative evolution.

interactions of the ingredients, depending to some extent on the environment, determine the eventual form. For a complex system, this principle predicts a huge multitude of forms, yet an important mystery of biology is that a very few forms are repeated over and over. Thus, we must recognize that the development of biological form is canalized to proceed in certain preferred pathways.

One of the thorns in the side of genetic determinism is that even small differences in the gene pool of a species can produce great differences in form. For example, between mice and humans, the gene pools are quite similar, but the difference in forms and functions is far out of proportion to the difference in genes. This evidence runs contrary to what we would expect from genetic determinism: The more similar the gene pool, the more similar the macroscopic form and functions should be. The organismic answer is that small genetic differences are amplified by the dynamics of form, producing vast differences in the final species.

So we are looking at a dynamics of form that is deterministic but extremely sensitive to the initial conditions, that is, the genetic variations and the environment. This kind of dynamics was unknown in classical physics until the discovery of chaos theory.

Chaotic systems exhibit patterns of change that are qualitatively similar but quantitatively a little unpredictable at any given time. Such a movement is said to happen under the spell of a "strange attractor." Small changes in the initial condition of the system may sometimes be enough to trigger a shift out of the old attractor "basin" to a different basin driven by a new attractor.

In the living world we also see the same characteristics in the forms of organisms belonging to the same species. Individuals of a species are qualitatively similar in form, but each individual form is quantitatively a little different from the rest. This heterogeneity of biological form could very well be the result of chaos dynamics operating in the system near attractor basins.

Here's the most important point: The change in initial conditions brought about by the quantum possibilities of new gene mutations may be enough to trigger a shift to a new attractor, a change that is certain to generate new form, new function, or both.

Recall that before a collapse, gene variations exist only as possibilities, and in this way different confluences of possibilities, when collapsed, will give rise to movement under the spell of different attractors in different basins. In this way, coupling the chaotic system to the quantum possibilities of gene variations introduces multiple pathways or modules of chaotic movement, each with its individual attractor and individual form and function. These modules of chaotic attractor dynamics are the macroscopic amplification of the microscopic gene variations.

Out of all these possible modular forms, consciousness chooses the gestalt that most closely resembles the vital blueprint that best represents a particular biological archetype of form (and function). The gestalt manifests into new form tangled hierarchically, in a discontinuous collapse event.

CREATIVITY AND COMPLEXITY: PROGRESSION IN EVOLUTION

We've looked at an explanation for how form can change in evolution, so we can now look at why it changes in the particular direction of increasing complexity. We can start from the adage that God created humans in His (or Her or Its) own image. By looking at our own creativity, we can gain insight as to how God's creativity in evolution works.

For example, I have always been fascinated by descriptions of the creative culture of the people of Bali. Authors portray a wonderfully innocent and "unsophisticated" way of life that is quite creative. But Balinese creativity, while prolific, also remains undeveloped and unsophisticated. That is, the Balinese create with great enthusiasm, but the forms they produce fall within a tightly constrained aesthetic.

Such a culture, although creative, doesn't evolve toward complexity. Cultural evolution opens the door for the discovery of more sophisticated concepts, such as scientific laws or the complex tapestry of a symphony. In turn, sophistication requires new creativity if the culture is to proceed, making for a ladder-like pattern of development. The contexts discovered earlier via fundamental creativity act as scaffolding for new creative acts of situational creativity, which lead to more complex discoveries and expressions later. Creativity itself evolves as the context of its invocation evolves.

What does this mean for biological creativity in evolution? Every act of biological creativity leading to a new trait and a new species is a new expression of the archetypal biological functions. The new trait opens up the possibility of the evolution of further new traits that represent either the same biological function or an entirely different biological function. Some traits work particularly well as scaffolding for new traits, giving evolution an enormous opportunity for proliferation of complexity. Others are like a blind alley; they do not allow much further creativity, and for them stasis is the rule.

We find exactly this pattern in the evolutionary data. Frogs have existed longer than mammals and exist in the form of thousands of species. Yet the anatomical similarities among them compel biologists to classify them all in one order. One type of frog has remained in stasis to such an extent that ninety-million-year-old fossils and present-day specimens are classified in the same genus, *Xenopus*. Then there are the mammals, a much younger group compared to the frogs, but one offering such an enormous scope for fundamental creativity that zoologists classify them in twenty-four orders. Mammalian traits have certainly provided raw material for complexity!

INTERMEDIATES REVISITED: EVIDENCE FOR THE SUPRAMENTAL

Whatever the level of complexity, the end result of biological creativity is a representation in physical reality of the archetypes of biological functions. As discussed in chapter 4, these archetypes exist in the realm of the supramental. Biological evolution gives one very interesting piece of evidence for the archetypal nature of biological functions, and thus for the supramental. To find it, we will return to the subject of intermediates.

Previously I argued that biological macroevolution proceeds through creative quantum leaps in which a species transforms into a novel new species by developing the capacity to represent a brand-new blueprint of the vital body of morphogenetic fields. The new representation is an organ or trait. These organs bring to being either a refinement of an already represented biological function or a manifestation of a biological function never represented before.

This aspect helps explain a curious feature of biological evolution. A certain trait initially comes to being to represent one biological function. Later, however, a more sophisticated trait comes into manifestation all of a sudden, clearly resembling the previous trait but performing an entirely different function. In other words, the same blueprint has been used, only in a more sophisticated way; the sophistication serves to express a quantum leap to a new biological function, a new archetype.

Darwinists, of course, try to get some mileage out of this feature of evolution in support of continuity. As discussed earlier in this chapter, however, the existence of intermediates does not necessarily prove continuity; instead, it may be striking evidence of discontinuous quantum leaps of creativity in biological evolution.

An example will make this clear. Darwinists make a big case for another intermediate, this time a series of them that allegedly arose between land mammals and cetaceans—swimming mammals. In the late 1980s and early 1990s, the biologist Phillip Gingerich and collaborators uncovered this series. The animal at the midpoint of the series was named *Ambulocetus natans*, "the swimming whale that walks." The shape of the front and hind limbs of the fossil remnant make it quite plausible that the animal could both walk and swim. The Darwinists assume that this new trait could come about continuously from gradual modification of the land mammal with walking limbs only. They forget that swimming requires many other internal modifications (including some involving the brain), all acting coherently as a whole, besides just the shape of the limbs!

In the theory of creative evolution, we accept the new trait as the quantum leap to express a new biological function—a new archetype—that of swimming. The emergence of the "walking whales" is an instance of fundamental creativity, the leap into the supramental realm. The change from these creatures to the earliest cetaceans, the archeocetes, occurred through situational creativity working within established archetypes, a process of refinement.

CATASTROPHES: SYNCHRONICITY AND THE ROLE OF THE ORGANISM

Some consensus has been reached that the dinosaur extinction some sixty-five million years ago was brought about by a large meteor shower (or, at least, by several of them). This event (or events) made room for the very important evolution of mammals that led eventually to the evolution of the human being. Evidence shows that mammals were already on the scene, but the catastrophe made room for their creative expansion.

So was the evolution of humans on Earth brought about by pure, meaningless chance events? If that is so, then how can we uphold purposiveness of biological evolution, when clearly God's purposiveness—to evolve life in the direction of humans—needed the contingency of chance events?

We need not see contradiction here with the scenario of biological creativity and purposiveness. Contingencies are often very important in the history of human creative acts, except that we recognize them as components of events that psychologist Carl Jung called "synchronicities." An event outside in the material arena (the meteor shower) and an event inside in the biological arena (the act of biological creativity) occur simultaneously and meaning and purpose emerge, the evolution of a new mammalian brain. This conjunction of events qualifies as synchronicity in Jung's definition (1971).

As catastrophe theorists point out, these synchronistic events are important because they open the evolutionary landscape for the newly created macro organisms. They also create a sense of survival urgency for evolution in the species that survive the catastrophe. A sudden change of environment requires an equally sudden evolutionary jump in the species. There is no time to wait for slow Darwinian evolution to bring adaptation.

Here is our opportunity to see the role of the organism in evolution, a role that Darwinism denies. All creative people know that human creativity requires motivation, an urgency that usually takes the form of a burning question and an intention to find a solution. From the point of view of the whole quantum consciousness, or God, the motivation is the need to move change in a certain direction, to drive evolution purposively. In the event of an environmental catastrophe, this evolutionary motivation resonates speedily with the individual organisms because it coincides with survival necessity.

This sounds more like Lamarckism. Lamarck is famous for ideas such as environmental sensitivity and intention of the organism. But there is now experimental evidence in favor of a kind of Lamarckism in the face of environmental upheaval.

In 1988, the biologists John Cairns and his collaborators demonstrated that a certain kind of bacteria, when proffered food that it cannot digest directly but can digest if it undergoes a one-step mutation, hastens its own mutation rate and survives. This phenomenon, which even the popular magazine *Newsweek* heralded as evidence for Lamarckian evolution, is now called *directed mutation*. Even in microevolution, then, Lamarckism sometimes persists. I submit that macroevolution under the influence of biological creativity is closer to Lamarckism than Darwinism.

I also suspect that biological organisms have nonlocal connections through feelings, through the experience of vital movements of the morphogenetic fields. Because of the dominance of the mind, this nonlocality has become shrouded for us humans. The rest of the biological world is, however, nonmental, or at least largely so; it is not limited in the same way. The nonlocal connection through the vital feelings acts as a species or even larger group consciousness (which can be seen as a generalized group ego; see chapter 12). I think that this group consciousness intends evolution in response to rapid environmental changes and that quantum consciousness/God responds to this call. The outcome is a quantum leap in evolution.

Does any direct evidence exist for nonlocality in evolution? Yes. In a phenomenon called *coevolution*, two entirely different species must evolve together to better survive. We can see coevolution most easily in symbiotic relationships. For example, many plants have a symbiotic relationship with ants. The plants provide the ants with shelter and food, and the ants drive away harmful insects and animals, and even help keep other vegetation away. Indeed, some plants independently evolve seeds with an edible fatty lobe as a reward for ants. The ants on their part reward the plants by carrying the seeds away to their homes, eating the fatty lobes, and leaving the seeds to grow in places that are favorable for the plants. The Darwinists argue that each species puts selection pressure on the random variations of the other as part of the other's environment. But this view misses the fact that the probability of simultaneously beneficial variations is extremely low. A much more probable answer is that the necessary mutually beneficial variations are held in the limbo of quantum possibility until the time is opportune for nonlocal consciousness to collapse the two correlated sets of variations simultaneously.

12

$\mathcal{R}$ECONCILING $\mathcal{D}$ARWINISM and $\mathcal{G}$AIA $\mathcal{T}$HEORY

We've seen in the previous chapter that natural selection cannot be the source of macroevolution. What then is its role? The biologist Ernst Meyr has called natural selection the sculptor of evolution; George Simpson likewise compared its role to that of a poet or a builder. These biologists were ready to attribute all the power of biological evolutionary changes to natural selection and adaptation, and here we are, taking that power away.

The truth is, adaptation is overrated as an evolutionary mechanism. The authors Augros and Stanciu (1988) put it well when they say, "Adaptation has lent itself to wide abuse." As the biologist Niko Tinbergen complains, "Some men went so far in supporting improbable theories about the survival value of organs, color patterns and behavior that they gradually discredited this whole line of research. One well-known and respected naturalist seriously claimed that the bright-pink coloration of the roseate spoonbill served to camouflage this bird at sunrise and sunset—without trying to consider how this bird managed the rest of the time."

Gould and Lewontin (1979) elaborate on the problem:

> We would not object so strenuously to the adaptationist pro-
> gramme if its invocation, in any particular case, could lead in
> principle to its rejection for want of evidence. . . . [I]f it could
> be dismissed after failing some explicit test, then alternatives
> would get their chance. Unfortunately, a common procedure
> among evolutionists does not allow such definable rejection for
> two reasons. First, the rejection of one adaptive story usually
> leads to its replacement by another, rather than to a suspicion
> that a different kind of explanation might be required. . . . And
> if a story is not immediately available, one could always plead
> temporary ignorance and trust that it will be forthcoming [this
> I call promissory Darwinism]. . . . Secondly, the criteria for the
> acceptance of a story are so loose that many pass without prop-
> er confirmation. Often, evolutionists use *consistency* with natu-
> ral selection as the sole criterion and consider their work done
> when they concoct a plausible story. But plausible stories can
> always be told. The key to historical research lies in devising
> criteria to identify proper explanations among the substantial
> set of plausible pathways to any modern result. (587–88, also
> quoted in Augros and Stanciu 1988, 194–95)

As I have suggested, adaptation is a species' way to arrive at
homeostasis—stability. Augros and Stanciu (1988) agree:

> What Darwin took for a source of evolutionary change in a
> species [i.e., adaptation] is, in fact, a source of stability. An indi-
> vidual plant, for example, may assume rather different forms
> according to soil conditions, winds, altitude and other external
> conditions. This adaptability helps it to make the best of many
> different habitats, but it does not carry the plant beyond the
> bounds of its genotype. In a similar way, individual variation or
> polymorphism [existence in more than one form] does not take
> a population beyond the bounds of the species. (195)

Do natural selection and adaptation play any role at all in evolution? The role is obvious if you remember the four stages of biological creativity: preparation, incubation, insight, and manifestation. To recap, preparation consists of creating the urgent necessity of creativity in consciousness, developing the burning question of survival. Incubation and insight we have discussed already. What is the manifestation stage of biological creativity in evolution?

It is simple. The new has evolved, but it has to adapt to the environment; otherwise it will not survive! This creative necessity of environmental adaptation is what Darwin recognized as the necessity of survival. Darwin believed in gradualism, so he viewed necessity of survival as driven by the selection pressure of environmental changes.

Instead, the right view is that the selection pressure of environmental changes acts in the preparation stage, contributing to the creative urgency in consciousness. The necessity of survival drives the later, manifestation stage of expression. The manifestation to adapt to the environment takes place either through situational creativity, through tracking the environment (Eldredge 1985), or even through Lamarckism (i.e., gathering more information and then making the necessary adjustment; see chapter 20).

The fossil record actually supports this new interpretation of the role of natural selection. If natural selection is the causal mechanism for adaptation, it is very hard to see how a species can be maladapted to the environment. In the new approach, maladaptation can be readily explained as an incompleteness of the creative insight that leads to the inadequacy of the evolved trait or the manifestation of the insight. In other words, maladaptation is a failure to find an appropriate situational adjustment or a niche that fits the trait. For an example of creative maladaptation in which a species develops a trait that does not quite fit the survival necessity, consider the stinging bee that ensures its own death by the very act of stinging. It makes no sense as a Darwinist adaptation, but perfect sense as an inadequate creative manifestation. Another example is the peacock's tail, which is counterproductive in terms of utility; natural selection should not have allowed it. But if evolution is the result of creative quantum leaps, then survival necessity is just one criterion on the creator's list and can be relaxed on occasion in favor of other criteria, such as beauty in the archetypal, objective sense.

DARWINISM AS
THE CORRESPONDENCE LIMIT OF
SITUATIONAL CREATIVITY

What happens between the leaps of quantum evolution? It is easy to suspect that a Darwinian slow mechanism should now be enough to cope with slow and gradual environmental changes. But there are subtleties.

First, note that a creative leap expresses a whole bunch of new genes. In some combination, these genes make specific organs. But a gene can be used and is used in more than one combination and in more than one context of living. The creative leaps of evolution cumulatively build up what is called the gene pool of the entire species; the adaptive needs of the species can now be met from this pool without developing new genes.

In human creativity, the ability to adapt to societal needs through the invention of new combinations of previously discovered ideas and contexts is called *situational creativity*. Situational creativity is the creativity of invention, as opposed to fundamental creativity, which is discovery. But here, too, there is a subtlety. Ask yourself, Can a computer generate a creative product via situational creativity? Well, you say, aren't there some examples of computer-generated art and poetry? You are right. Consider, for example, the following computer-generated poem:

There was a time when moorhens in the west
There was a time when daylight on the top
There was a time when God was not a question
There was a time when poets
Then I came. (Boden 1990, 1)

This "poem" was "written" by the computer program called *Arthur* (an acronym for *a*utomatic *r*ecord *t*abulator but *h*euristically *u*nreliable *r*easoner). You may recognize that some lines originated in the minds of famous poets; Arthur took these known contexts and put them together in a new context of abstract meaning that sounds intriguing and can easily pass for a (situationally) creative poem. The subtlety lies in this: Arthur might have compiled the poem, but Arthur could not have recognized its meaning value. It takes a human being to do that. Thus, consciousness seems to be needed even for situational creativity (Goswami 1999).

The beauty in the case of situational creativity in Darwinian gradual evolution is that natural selection and adaptation to the environment do the job of recognizing the creative accomplishment, and consciousness is not needed. Thus, Darwinism can be regarded as the "correspondence limit" of creative evolution. Any new scientific paradigm must incorporate the old paradigm in some limit. This is called the *correspondence principle*, which was formulated by the great physicist Niels Bohr.

A good example of Darwinian evolution is the famous case of gypsy moths in London, in which the moth population changed color from brown to black in response to heavy air pollution. The gene pool already contained the "black gene." The individual moths in which the "black gene" was expressed gained selective advantage over the moths of the old color in the pollution-darkened city. Hence they survived, whereas the others didn't. Rapidly, natural selection eliminated the moths of the old brown color in favor of the new black ones.

Finally, as Gould and others have noted, the fossil data also show vast epochs of virtual stasis in the evolutionary history of many species. This stasis is a serious embarrassment for classical Darwinism. But for creative evolutionary thinking such phases correspond to our own ego-homeostasis, the limit of conditioned existence when no creativity, situational or fundamental, is needed to deal with environmental changes.

WHO CREATES? THE DANCE OF GAIA IN EVOLUTION

One question I have left largely unaddressed in all this discussion: Who creates? Of course, in the ultimate sense, in any creative act the creator is the downward causation from quantum consciousness/God. But in truth, that statement is strictly true only for the quantum stages — incubation (or unconscious processing) and insight. For preparation and manifestation, a manifest consciousness has to be involved. Quantum consciousness won't do, because it is always transcendent, not manifest. So who creates? It is a tangled play of the quantum consciousness and a manifest consciousness. (The tangledness explains why the creation can be and often is imperfect.)

What is the nature of this manifest consciousness? For human mental creativity, it is the individual ego that prepares, that manifests the insight gained with quantum leaping, that plays with quantum consciousness in alternate doing and being (a process that I sometimes call do-be-do-be-do). In the biological creativity involved in evolution, the creative transformation occurs at the level of the species, never in the individual. So the manifest consciousness is at least the identity of consciousness with an entire species—species consciousness.

But sometimes the specific change of trait(s) that takes place in a quantum leap is so fundamental that it becomes the beginning of a new genus, family, order, class, phylum, or even kingdom. In that case, the manifest consciousness involved must be the conscious identity of the entire new group involved.

There is a more fundamental way of looking at all this. Evolution begins from one living cell, the first that sets the whole ball rolling. Life could not have originated twice; with this all biologists agree. The one unnucleated cell becomes nucleated, then becomes multicellular, branching into the three kingdoms—fungi, plants, and animals. Each of these transformations is a gigantic quantum leap of creativity. In the animal kingdom, creative transformation brings first the invertebrates, then the vertebrates, beginning with fish. From the fish come the amphibians, and then the reptiles. The reptiles then quantum leap to the two branches of birds and mammals.

We see, in a sense, a single life evolving in many branches, transforming itself through many quantum leaps and at the same time maintaining itself in the many different states of homeostasis that we witness in Earth's biosphere, its biota. The quantum consciousness identified with this one evolving life on Earth I call Gaia-consciousness.

James Lovelock's Gaia hypothesis (Lovelock 1982) has had its followers and opponents. The idea came from Lovelock's observation that the Earth, through its history, has maintained more or less the same average temperature even though the sun is hotter. He proposed that Earth itself acts as a living system maintaining its homeostasis. Later the biologist Lynn Margulis supported the idea. As mentioned before, Margulis' own work on the transition from prokaryotes to eukaryotes suggested the importance of symbiotic relationships in evolution. To Margulis, the Gaia hypothesis reflects the symbiotic relationship of the living and the nonliving on our planet.

But the Gaia hypothesis is much misunderstood. Many of Lovelock's followers think that he is talking about the whole planet literally as a living organism, and, indeed, Lovelock himself sometimes creates that impression in his writing. One biologist (see Harman and Sahtouris 1998) has even poetically generalized the idea to think of a star system, a galaxy, and, indeed, the entire universe itself as alive! Later writings by both Lovelock and Margulis make it very clear that they don't mean that the inanimate part of Earth is anything like a living object. By *Gaia* they mean the living biosphere (Lovelock's word is *biota*) acting as a whole.

This interpretation fits with my assertion above. And when we recognize that in the quantum manifestation process of life, one consciousness divides itself into life and environment, the living and nonliving on Earth, the symbiotic relationship of the pair becomes obvious. The nonliving is not outside consciousness. In this way, we can think of the nonliving Earth's thermostatic control and the living biosphere's creative evolution and homeostatic survival as events of synchronicity, their common cause residing in quantum consciousness itself.

As we look at biological creativity in evolution as Gaia-consciousness taking quantum leaps during its dance of life, choosing its various living forms in the journey toward increasing complexity as a means to achieve better and better self-expression, a different take on Darwin's insight of natural selection and on Lovelock's insight on Gaia begins to emerge.

Fecundity and adaptation, according to the biophysicists, define Darwin's theory (see chapter 8). But Darwin himself was quite aware of the importance of fecundity and adaptation; he could have defined his theory that way in the first place and saved us much agony from controversy and politics. Why use the phrase "natural selection?" The usual interpretation of the word *selection* is that individuals of a species compete and the fittest is selected (by nature through adaptation). And what is nature? Darwin and today's biologists, being under a materialist influence, tend to think of nature as inanimate. A selection by the inanimate is not choice; it is mechanical.

In Gaia theory the opposite is emphasized. There, the living part of nature, Gaia, regulates the nonliving environment in the process of maintaining homeostasis for the biota.

But surely nature must mean everything that is manifest, everything that is the case. In other words, nature on Earth is both the inanimate part of nature, the environment that Darwin emphasizes, and the biota, the Gaia of Lovelock. Can we use the word *selection* to mean "choice," not of the fittest individual, but of the best creative possibilities of the vital that will best express consciousness? Natural selection then becomes the choice that Gaia makes symbiotically with the nonliving environment for their joint evolutionary journey of making the unconscious conscious. In truth, the environment doesn't select the biota, nor does Gaia—the biota chooses the environment, but, from behind the scenes, quantum consciousness/God chooses the course of both. In this way, I would like to think that Darwinism and Gaia theory together complementarily express the complete truth about evolution.

What Did Darwin Mean by Natural Selection?

Biologists' reduction of teleology to teleonomy through a combination of Darwinism and genetic determinism is logically inconsistent if we insist on using the original purposive statement based on survival: Darwinism requires survival (for its formulation), but survival itself requires Darwinian adaptation. And if, following the recent work by biophysicists, we reduce Darwinism to a purely physicochemical, nonteleological statement—the organisms that reproduce best, adapt best to the environment—then Darwinism can no longer be expected to explain program-like behavior.

In quantum thinking, genetic determinism gives only part of the answer—the possible variations. However, natural selection in

Darwinian form cannot collapse these possibilities into an actual change; that requires consciousness. But if we reinterpret "natural selection" as choice by nature in the form of Gaia-consciousness according to the creative requirements of the situation, this selection *can* collapse the possibilities into actuality.

The first part of Darwin's theory, that explaining variation, remains intact, but natural selection is interpreted differently. The lesson is that a great scientist's creative insights can bear fruit in more than one way, though we may have to wait a long time to gather their true meaning!

Perhaps another example from physics will convince the biologist of the plausibility of the argument presented here. It is well known that Einstein spent his last year working on a unified field theory, but he got stuck trying to unify gravity and electromagnetism. Physicists eventually made a breakthrough in unifying the different force fields, but the breakthrough depended on first unifying electromagnetism with the weak nuclear force.

Same here. The breakthrough in developing a complete theory of evolution is coming through an expanded meaning of the concept of both nature and selection.

DARWINISM OR CREATIVE EVOLUTION?

Evolution is! But is evolution synonymous with Darwinism, or is there another theory that better satisfies the theoretical demand for logical consistency as well as the empirical demand for agreement with data? In the last three chapters, we have discussed the trials and tribulations of Darwinism throughout its one-hundred - fifty-year history. We also have looked at how creative evolution unambiguously explains some of the major data for evolution for which Darwinism seems an ad hoc and ambiguous explanation, at best. In this section, I present additional data in support of creative evolution.

One characteristic of fundamental creativity is that when exploration begins and there is an open vista of ideas to discover, the quantum leaps tend to be gigantic. As ideas are explored, the fields of individual creative exploration become narrow. In science, it is much more fun to explore a new field than a field for which there are many known contexts and where creative exploration has become canalized.

In biological creativity, creative evolution would then predict that the farther we go back in geological time, the more astounding the cases of creative leaps in speciation should be, the kind of leaps that give rise to many new contexts (i.e., biological groups such as phyla) for further evolution. With time, there should be a corresponding reduction in the evolution of new contexts. Eventually, there would only be the narrowest of contexts for change—species to species within the same genus.

What do the data say? According to the physicist Gerald Schroeder (1997),

> ... until the mid-1980's the understanding of the development of animal life was that it had followed the logical path of a gradual evolution with more simple phyla over eons leading into more complex phyla. With the rediscovery of fossils held quietly in the drawers of the Smithsonian Institution since 1909, this concept underwent a drastic revision. These fossils in conjunction with other discoveries indicate that all animal phyla appeared almost simultaneously 530 million years ago in the Cambrian period. All further development was confined to variations within each phylum. One of the great mysteries of animal evolution is why no new phyla have appeared since that Cambrian explosion of life. (88)

The zoologist James Brough points out the same thing, that there have not been any new phyla since the Cambrian age, five hundred million years ago. And new classes within a phylum stopped emerging some four hundred million years ago, since the lower Paleozoic era. The next lower hierarchy, new orders, stopped emerging about sixty million years ago, at the end of the Mesozoic era. Says Brough (1958),

> Evolution seems to have worked in a series of more and more restricted fields with large scale effects steadily decreasing . . . as to the future, evolution may go on working in smaller and smaller fields until it ceases altogether.

This trend is exactly what creative evolution would predict, except, of course, that when the evolution has just about ceased in vital representation making, evolution of the representation making of the mind takes off.

Gould has called this the "pattern of shift from few species in many groups to many species in fewer groups." Darwinism favors the opposite trend. If evolution is the accumulation of small variations, we should expect orders to appear only over long periods in time, new classes to take even longer periods, and new phyla the longest time of all.

The maverick biologist Richard Goldschmidt thought along the same lines that I propose. It is the phylum that contains classes, he said, and asks:

> Can this mean anything but that the type of the phylum was evolved first and later separated into the type of classes, then into orders, and so on down the line? This natural naïve interpretation of the existing hierarchy of forms actually agrees with the historical facts furnished in paleontology. The phyla existing today can be followed the furthest back into remote geological time. Classes are a little younger, still younger are the orders, and so on until we come to the recent species which appear only in the latest geological epochs.

What is the Darwinist's reply? Darwinist guru Ernst Mayr lamely says the categories (phyla, etc.) are man-made artifacts.

In contrast, biological creativity supports both Goldschmidt's logic and paleontological fact. All fields of human creative endeavor have one thing in common: Creativity in a new field takes off with an enormous flare. With time and proliferation of contexts (subfields), complexity develops and the speed of progress slows.

The same pattern is observed in biological evolution, giving further credence to the creativity theory of evolution. For the fossil data, sudden new bursts of flora and fauna are called *radiation*. During the early Cenozoic era fifty million years ago, some time after the extinction of the dinosaurs, mammals suddenly exhibited an amazing radiation into about twenty-four different orders—rabbits, rodents, elephants, cetaceans, and primates included. The resulting Age of Mammals took only twelve million years to establish itself. After that the progress slowed down.

Another important facet of creative evolution shows up in the phenomenon of parallel evolution, the most famous case of which is the evolution of placental and marsupial mammals. The similarities of form—body plans—of corresponding animals of the two groups are remarkable. In the Darwinism-plus-geographical isolation model, the genes, although originating from a common ancestor, would diversify rapidly, changing the form as well. But in the creative scenario, the forms are imposed by the morphogenetic blueprints that remain the same for both parallel lines, accounting for the persistence of the similarities of form.

PART 4

FORM, FEELING, and VITAL ENERGIES

CHAPTER

13

$\mathcal{M}$ORPHOGENETIC $\mathcal{F}$IELDS, $\mathcal{E}$VOLUTION, *and the* $\mathcal{D}$EVELOPMENT *of* $\mathcal{F}$ORM

A major component of the theory of creative evolution is the postulate of the vital body, which is the abode of the morphogenetic fields—the blueprints of biological form. In the nineteenth century, vitalism became popular as a source of a "formative force" needed to create biological form from an embryo. In contrast, biologists of the materialist ilk claim, more or less universally, that molecular biology has sufficient explanatory power for everything connected with the life of a cell, including its embryonic development. They take this step by adding one more component, information contained in the genes, to the prevailing concept of a living biochemical machine of matter and energy.

To use the familiar language of the computer, the part of a biosystem carrying hereditary instructions (called the *genotype*) is the software, a deposit of instructions, and it is assumed to be capable of building form, the biological hardware (called the *phenotype*). No vital formative force is deemed necessary in this way of thinking.

The way we look at it in this book, the program-like behavior of biological forms (including the genes) is not the result of chance and necessity, and it is not restricted to the genetic level. Consciousness, through the use of morphogenetic blueprints, programs both the genotype and the phenotype. The actions of the genotype programs, the genetic code and the regulator genes, are easy to see and hard to refute, and the materialists include them in their theories, the unsolved mystery of their origin notwithstanding. The program-like, purposive activity at the somatic level is a little more subtle; materialists fight tooth and nail to explain away this activity as mere teleonomy.

But detailed analysis shows the inadequacy of such attempts to explain away program-like behavior at the macro, phenotype level (Behe 1996). Basically, for any complex biological function involving one or more organs, we find a series of purposive components acting upon one another in chronological order in the same manner that programs act on programs in computer software to bring about a purposive function. The chronological order is maintained because these material-level programmed components are correlated with vital-level morphogenetic fields that consciousness can feel. This feeling serves as a guide: In ordinary computer software, program-like behavior is guided by human thinking; in biological organs, program-like behavior is guided by feeling. In this way, the morphogenetic fields give us a profound explanation of feeling: what we feel, how we feel, and where we feel.

But is there independent evidence for the existence of morphogenetic fields? Yes, certainly. I will just touch on the main threads here, leaving detailed discussion for later chapters. The first source of evidence is that we can measure the vital energy we feel by measuring its material correlates (see chapter 14). Further evidence for the vital body arises from its importance in systems of medicine developed not only in the East in the form of traditional Chinese and Indian medicine (the latter is called *ayurveda*) but also in the West in the form of homeopathy. This aspect of the vital body is further discussed in chapter 14. There is, additionally, a third important piece of evidence for the morphogenetic fields. A most welcome change in biology today is a partial revival of Lamarckism. However, although the data are there, theory has lagged behind. I will show in chapter 20 that the idea of morphogenetic field enables us to understand why and how Lamarckism works as an evolutionary mechanism. In the process, we will find an explanation of the important but hitherto unexplained phenomenon of instincts.

Before looking more closely at such evidence, however, I will in the remainder of this chapter lay additional foundation on the nature of the vital body and its morphogenetic fields.

THE VITAL BODY AND ITS MORPHOGENETIC FIELDS REVISITED

In the nineteenth century and even in the early twentieth century, the vital body played an essential part in biological thinking. For example, the much-admired philosopher Henry Bergson (1949) saw life as an expression of élan vital—vital essence, the special feel of life from inside, what we today call vital energy. Things changed drastically when molecular biology provided a picture of a cell containing DNA for replication and proteins for various functions of maintenance. This picture of the cell seemed to have all the explanatory ingredients for biological functioning. Molecular biology and Darwin's theory of evolution packaged under a new synthesis called neo-Darwinism together became the new paradigm of biology. The concept of vital energy was considered extra baggage and abandoned. It was thought that vitalism smacked of dualism besides, and no scientist wants to be tarred with that brush. So good riddance!

By 1960, though, biologists like Conrad Waddington (1957) were already pointing to a problem that might be beyond the scope of the new paradigm: the problem of biological form making, or, technically, morphogenesis—how form is created from a single-celled embryo. This cell makes more of itself by cell division, in the process creating an exact replica of itself, with exactly the same DNA. But then why does a cell belonging to the heart behave so differently from a brain cell? How do the cells belonging to different organs get so differentiated?

The cells of each type of organ function differently because they make different proteins by activating different sets of genes. The source of the program that determines which genes to activate is called a morphogenetic field. Nonphysical and nonlocal, a morphogenetic field is the blueprint of form, an idea reintroduced in biology by Rupert Sheldrake (see chapter 4). The vital body is the reservoir of these fields. However, we must recognize that the morphogenetic fields do more than just help create programs for regulating genes. They also help with the dynamics at the phenotype level.

This new view—seeing morphogenetic fields as the source of cell differentiation—can integrate two different views of form making currently prevalent in biology: the establishment (gene-based) view and the organismic view.

The establishment view is based on genetic determinism. The genes themselves, the DNA, have all the programs for form making. That is, the genotype determines the phenotype, except for a few details of the developmental process.

But there is also a minority view of development that I call the organismic view. According to the organismic view, morphogenesis is not the result of genetic programs. Form is the result of a dialectic relationship: The underlying dynamics (structured by the genes) generate the geometry of the form; the geometry modifies and constrains the dynamics. The genes are secondary; they only establish the range of the parameters within which the dynamics of form play out under the guidance of "morphogenetic fields" (a different sense of the term, as we will see in a moment). The biologist Brian Goodwin (1994) gives an example from the world of high plants—plants with roots, stems, leaves, and flowers—of which there are a quarter of a million species. Goodwin points out that despite the apparent individuality of leaf shapes, these shapes fall into only three basic patterns: whorls, alternating, and spiral. The same applies to the flowers of these plants. This fact suggests, according to Goodwin, that there are only three basic morphogenetic attractors (the language of chaos theory again) for the dynamic organization of the growing tip of the plant, a region called the *meristem*.

I agree with Goodwin except for the naming of the concepts. I agree that the self-organizational dynamics of the developing forms drives the form toward the attractor; but these fields are entirely local and therefore should not be termed morphogenetic fields. Following Rupert Sheldrake, I reserve the concept of morphogenetic fields for the situation when nonlocality enters form making. The local dynamics produces macroscopic possibility forms from which consciousness may choose; this choice and the resultant collapse of possibility forms is nonlocal; and for this choice consciousness needs the morphogenetic blueprints.

In simple terms, consciousness acts in the manner of an architect building a house. The architect starts with the basic building material and skillfully guides the internal dynamics of the material to make form according to the blueprint of the house. The same is true for living forms, except that now everything, the blueprint as well as the building material, is a wave of possibility. The available forms are determined to a large extent by the genes, which are themselves possibility forms. The self-organizational dynamics of the "possibility stuff" of the living form plays out, producing many macroscopic possibility forms (e.g., organs) produced by many attractors. When quantum consciousness sees a match (i.e., a resonance) of movements between a morphogenetic blueprint and a gestalt of many possibility patterns of form for an organ, it collapses actuality. A similar process results in the development of a function involving several organs.

Let us be clear about what the final actualities are. The physical actuality is the form, the shape and function of the organ (or organs)—this the biologist acknowledges and everybody can verify. But a second component is present: the manifest morphogenetic field, present in awareness in the psyche of the living being and experienced as feeling. This internal feel is the feeling of being alive that Bergson called "élan vital."

Note also that the form, the organs, are initially made in God-consciousness, but when the process of form making is over and we begin using the forms, our experience of feeling alive reflects more and more the effects of conditioning.

CREATIVE EVOLUTION
AND DEVELOPMENT

To return to the analogy, consciousness uses the morphogenetic fields to build biological forms as an architect uses a blueprint to build a house. This analogy can be taken further. An architect's blueprint is modular, as are the physical forms that are put together to represent the blueprints. In the same manner, in building a complex biological form with complex functions, morphogenetic fields are used to assemble modular forms. These individual biological modules are the result of chaos dynamics of macromolecules. An architect connects the physical modules through visualization and thinking. Likewise, in connecting the modules of biological forms, consciousness is guided by the feelings of the corresponding correlated morphogenetic fields.

The biologist Michael Behe correctly criticizes the Darwinists when he points out the biochemical challenge—impossibility, rather—of producing form that has a complex function or structure through Darwinian, step-by-step, linear mechanisms of chance and necessity. In form building, Behe points out, we repeatedly encounter what he calls "irreducible complexities." I have to agree, except I would use the concept of tangled hierarchies to make the same point: Forms cannot be put together bit by bit, modular though they may be. There is always a tangled hierarchy in the path of form building from micro to macro.

Behe (1996) gives an excellent example, describing the clotting of blood. In blood clotting, cascades of proteins appear with various functions. Some of the proteins in the cascade appear before the agent that is required to activate them appears. The apparent disconnect is much worse than the hypercycles that Eigen postulated for explaining the appearance of the protein-DNA duo in the living cell (see chapter 6). As Behe puts it, "the bottom line is that clusters of proteins have to be inserted all at once into the cascade. This can be done only by postulating a 'hopeful monster' who luckily gets all the proteins at once or by the guidance of an intelligent agent."

No doubt about it: The machinery involved in blood clotting is not the result of a piecemeal approach, nor does it involve a hopeful monster. Instead, the machinery is tangled hierarchical (irreducibly complex in Behe's language), and a discontinuous event of collapse is necessary for its creative formation; the entire gestalt of possibilities must be collapsed at once via downward causation. Because the final form is tangled hierarchical, it cannot be synthesized in simple step-by-step manner through anything like gradual Darwinian evolution.

More recently, some Darwinists (Carroll 2005; Kirschner and Gerhart 2005), to their credit, have attempted to take into account the role of development, or form making, in Darwinian evolution. But even these "evo-devo" approaches are inadequate because they make a fundamental assumption of continuity. As we have seen, this assumption is fundamentally wrong. Consciousness unawarely (that is, in the unconscious mode) processes the many possible modules of form until it finds a match between a certain gestalt of them and the blueprint for the organ or trait it is trying to evolve. The recognition of the gestalt leads to choice and discontinuous collapse, the quantum leap. The final form is tangled hierarchical and cannot be reduced to its parts and synthesized from the parts in any continuous manner. Because creative evolution already takes development into account, it is an evo-devo theory, but one freed from the shackles of Darwinian chance and necessity.

STABILITY OF FORMS

Ernst Mayr was a staunch believer that microevolution on a large scale leads to macroevolution, and he tried to prove it with selective breeding experiments on *Drosophila* (fruit flies). He bred successive generations of fruit flies, attempting to increase or decrease the number of bristles the flies grew, the normal average being 36. He reached an upper limit of 56 after 20 generations and a lower limit of 25 bristles after 30 generations. When the limits are reached, the flies die out.

Subsequently, he brought the fruit flies back to normal non-selective breeding and let nature take its course (Mayr 1963). Amazingly, the bristle count came back almost to the average in a mere five generations. This experiment is usually interpreted as an exhibition of genetic resistance to change and even given the label of genetic homeostasis.

I take a different view. I think this experiment illustrates a stability not only of genetics but also of form. The form of *Drosophila* fluctuates around the standard of 36 bristles; that form is what the original morphogenetic blueprint of the species indicates. So when nature is given a chance to take its course, the form always comes back to what the original blueprint indicates.

REGENERATION

More evidence for the role of morphogenetic fields in development comes from a fascinating case of regeneration, demonstrated in a series of experiments with flies. In these experiments, the researchers paired mutant genes to produce eyeless flies. Then they interbred these flies, and, indeed, the offspring were also eyeless. After a few generations, however, a few flies began to hatch with eyes—the form of the eye regenerated. The materialist explanation is that the genetic code of the flies includes a repair mechanism. But what activated the repair mechanism? A likely explanation is that with the help of the vital blueprint, consciousness is able to repair the mutant genes and regenerate the eyes.

Regeneration is, of course, a well-known phenomenon. A flatworm can be severed in many pieces, but each piece can regenerate and grow into a new flatworm. Another well-known case is the hydra's ability to grow back its tentacles. Even in us humans, the healing capacity of broken bones and severed nerves exemplifies regeneration. In each of these cases, the explanation is that with the help of the vital blueprints, consciousness is able to rebuild form.

MORPHOGENETIC FIELDS
AND THE BRAIN

Not so long ago, neuroscientists believed in the permanence of brain cells. Now, universally, there is excitement about brain plasticity. For example, learning changes the brain cells because new synaptic connections are made among them. But how does this happen? If you say that something triggers the genes to make new neural connections, think again. Learning happens at the somatic level; how does somatic information propagate to the genes without violating the central dogma of molecular biology? The answer is clear when we introduce the vital blueprints of the morphogenetic fields. These blueprints are available for making form whenever needed.

14

A NEW BIOLOGY
of FEELING

*E*volutionary biologists tend to think that everything about organisms — their structure, function, and behavior (especially reproduction) — is about evolution. But life is about living, and reproduction, though a major part of evolution, is only one part of life.

Molecular biologists, in turn, concentrate on the functional molecules of life—proteins and such. Of necessity, they don't study life while the living goes on. In this way, they too miss much about both life and living.

How do we study life without excluding living? Studying behavior is one way, to be sure, but that is only the outer aspect of living. There is also the inner aspect—feelings. These we can study only from the inside. Materialist biology, by ignoring the distinction of living and nonliving, also ignores the distinction of outer and inner in living, eventually ending up ignoring the inner aspect of living altogether.

FEELING:
THE INNER ASPECT OF LIVING

If you watch the television show *Star Trek: The Next Generation*, you may be impressed how its android character Data struggles to incorporate emotions in his being. Alas! The writers of *Star Trek* eventually gave in to the simplistic notion of an emotion chip. Unfortunately, most scientific thinking about this inner aspect of living is nearly as simplistic. Most establishment biologists think that emotion is a brain phenomenon, though researchers currently debate about the extent to which emotion can be understood in terms of neurology. Many biologists use that ultimate black box, Darwinism, when they need to say something useful about emotion.

Some biologists, to their credit, admit that feeling does not originate in the neocortex, that it is not computable. They even agree that feeling may occur before the cognition of thought during an emotion.

Sometimes, it is hypothesized that feeling and emotions are the territory of the neurochemistry of the limbic brain. To this end, the researcher Candace Pert's experiments on the "molecules of emotion" are important (Pert 1997). Certainly they are telling us something.

The more one investigates the materialist research, however, the more confused one gets. Isn't it our experience that feelings originate in the body? Otherwise, how would our culture perpetuate such notions as "butterflies in the stomach" or "heart warming" when talking about feelings? Only William James, the father of American psychology, had the right idea, and one that agrees with people's experience. According to James, feelings are associated with direct bodily changes when confronted with some stimulus. Then again, James was hardly a materialist.

Fortunately, in the psychology of the East, feelings are recognized as being associated with the physical organs, and emotions are clearly seen as effects of feelings on the mind and the body physiology. According to the Easterners, the body has seven major centers—the chakras—where we feel our feelings in a big way. Through the centuries, although the idea of the chakras has been empirically validated by practitioners from many spiritual disciplines, not much theoretical understanding has been developed. Now, finally, with the idea of Sheldrake's morphogenetic field as reinterpreted here, an explanation of the chakras—that is, where feelings originate and why—can be given.

MORPHOGENETIC FIELDS AND THE CHAKRAS

I have treated this subject in some detail elsewhere (Goswami 2004), so I will be succinct here. You can discover for yourself what a little quantum thinking enables us to scientifically theorize. First, look at the major chakras (fig. 12) and notice that each is located near one or more major organs. Second, note the kinds of feeling you have experienced at each of these chakras; use your memory of past feelings. Third, realize that feelings are your experiences of the vital energy—the movements of your morphogenetic fields. However, the same morphogenetic fields are correlated with the organs of which they are the blueprint and source. Now arrive at the inevitable conclusion: Chakras are those points of our physical body where consciousness simultaneously collapses the movements of important morphogenetic fields along with the organs of our body that represent these morphogenetic fields. Now was that so hard?

Before we go into case-by-case details of each chakra, it may be of interest for you to know the literal meaning of the Sanskrit word *chakra*. It means "wheel," a circularity, and thus is implicitly a reminder that a tangled hierarchical quantum collapse assures the arousal of self-reference at each of the chakra points. Our new science is validating ancient intuitional wisdom.

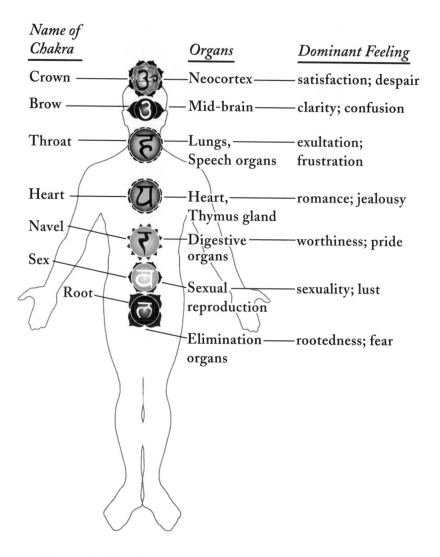

Name of Chakra	Organs	Dominant Feeling
Crown	Neocortex	satisfaction; despair
Brow	Mid-brain	clarity; confusion
Throat	Lungs, Speech organs	exultation; frustration
Heart	Heart, Thymus gland	romance; jealousy
Navel	Digestive organs	worthiness; pride
Sex		
Root	Sexual reproduction	sexuality; lust
	Elimination organs	rootedness; fear

Figure 12. The chakras and their associated organs and feelings.

Here is a chakra-by-chakra description of the vital function, the corresponding physical organs that represent the vital blueprints of the function, and the associated emotions and feelings at each chakra:

Root chakra. The vital function is waste elimination. The organs that express the vital blueprints of the function are the kidneys, the bladder, and the large intestine (rectum and anus). The associated feelings are self-centered rootedness, survival-oriented competitiveness, and fear.

I want you to notice from your own experience something important. When you feel rooted or competitive, the experience is a positive one for your emotional ego. We express this by saying that energy is moving into your root chakra. Notice the language used; it is not that physical energy is moving into the physical organs at the root chakra. Instead, the language signifies that vital movement (energy) is being generated in the vital blueprints of the organs via quantum collapse. On the other hand, when you experience fear, it depletes your emotional ego, signaling that energy is moving out of this chakra, the movement or change in the morphogenetic field is coming to a stop because collapse is taking place elsewhere; however much you try, you are unable to hold your attention at the chakra. Is this making experiential sense? These movements are not that unfamiliar, are they?

Sex chakra. The biological function is reproduction, its blueprints represented by the reproductive organs—uterus, ovaries, prostate, testes, etc. The feelings are sexuality and amorousness when vital energy moves into the chakra and unfulfilled lust when energy moves out and the chakra is depleted. This jibes with your experience? Why do many Americans watch so much sex or violence on the movie screen? Americans today live a very head-centered life. The vital energy movement is always up there at the top of the head, at the crown chakra. Could it be that the vicarious sex and violence of the media help to bring the energy back down to the sex chakra and the root chakra of competitiveness?

Navel chakra. The vital function is maintenance and the organ representations are the stomach, small intestine, liver, gall bladder, and pancreas. When energy moves into this chakra, you feel pride; when energy moves out, you feel unworthiness. You are unsure of yourself—there are butterflies in your stomach. Does this ring a bell? Here is something interesting: If energy moves into the navel from the sex chakra (unfulfilled lust or desire), we feel anger.

Heart chakra. The vital function is self-distinction (between me and not me). The organ representations are the heart and most importantly the thymus gland. The thymus gland is the crucial component of the immune system whose job is to distinguish those molecules belonging to my body (me) and those not belonging (not me), and to get rid of intruders. We feel romance when energy of this chakra is boosted. If the energy is depleted, the feelings are jealousy, loss, grief, and hurt.

Now this one should ring a bell in everyone; who doesn't remember a teenage *heart*throb? When we are older, the romantic feelings abate a little and are experienced as tingles or just warmth. The explanation is simply that your immune system has accepted the intrusion of your romantic partner into the vital field; in other words, as part of "you." When that expansion of who you are happens, it is possible for you to give anything to your partner, even your life. This is why we enjoy James Bond, who feels so romantic for a particular partner (though for only about six weeks) that he takes undue risks to save his partner's life.

When energy moves into the heart from the lower chakras, whether from the sex chakra after sex and the navel chakra after a good meal, you feel giving, you feel vulnerable to your own generosity. Men especially feel this effect, and many men suppress heart energy because they don't want to become vulnerable to their partners.

Here is something else interesting. When energy moves out of the heart into the navel when you are already angry because of unfulfilled desires, you feel hostility.

Throat chakra. The biological function is self-expression, and the organs that represent it are the lungs, throat, and speech organs; the organs for hearing; and the thyroid gland. When energy moves into this chakra, we feel the exultation for our freedom of expression. We feel the opposite, frustration, when energy moves out. This, too, should be familiar experience. Our throat dries up when we are unable to express ourselves.

Brow chakra (third eye). The biological function is evolution, and the organs that represent it are the midbrain and the hindbrain. The evolutionary impulse to develop the neocortex—mind mapping—must have been experienced at this chakra as an intuition. Even now this chakra remains the chakra of intuitive energy. You will notice slight feelings of warmth whenever your creative intuitional energies are high and you have clarity.

Conversely, when energy moves out of this chakra, we feel confused. I make the point later in this book (chapters 17 and 21) that right now our evolution in consciousness is blocked. Naturally, the entirety of humankind is confused. You will notice, if you pay attention, that there is today a general ennui, a lack of vitality, in this chakra.

Crown chakra. The vital function is self-knowledge. The organs are the neocortex, of course, where the mind is mapped, and also the pineal gland. When energy moves into this chakra, producing excess, we feel satisfaction; if energy moves out, we feel despair.

When discussing philosophy, when you are deep into metaphysical thinking about God, you will notice deep feelings of satisfaction. In contrast, consider existential philosophy or deconstructionism; clever though they are, engaging these ideas just doesn't satisfy. Instead, you will experience existential despair.

The materialists have it all reversed. They think we feel emotions in the brain, that is, emotions are brain epiphenomena that come to the body through the nervous system and the so-called molecules of emotion. Actually, it is the other way around. We feel feelings at the chakra first, then the control goes to the midbrain for integration through the nervous system and the molecules of emotion (the role of the amygdala is important here), and eventually the neocortex gets into the game when mind gives meaning to the feelings.

But where are the objective data on all this? Chakras are fun to experience, you say, but does any solid experimental data exist proving their importance, and hence the importance of the movement of vital energy? Yes, such data exist. This is the subject of chakra medicine.

Chakra Medicine

The fundamental idea of chakra medicine is that, for good health, the vital energy should be experienced in a balanced manner—it should move both in and out, without producing any ongoing excesses or deficits of quantum collapse, in all of the chakras. If the movement of vital energy at any chakra is imbalanced or blocked (suppressed), the corresponding organ or organs will malfunction and eventually become diseased. Healing of a chakra consists of restoring balance to the energy flow at the chakra. Physicians who have been engaged in chakra medicine, such as Christine Page (1992), have accumulated quite a bit of data and many success stories about the power of this healing method (also see Eden 1999).

An interesting aspect of chakra medicine is that imbalances are often produced by the mind. For example, the mental suppression of heart chakra energy, especially among American males, may be responsible for the malfunction of the immune system that causes cancer. Since mental processing is involved, this research is discussed in more detail in chapter 15, which focuses on issues of mind.

Robotomorphism or
Anthropomorphism?

Materialists, and most establishment biologists fall in this category, assume that animals are robots, a collection of programmed molecules, the software having been written through evolution by natural selection. This is a clear instance of robotomorphism (the term is from Augros and Stanciu 1988), but materialists claim it is the only possible objective view.

Of course, evolution also dictates that we humans evolved from animals, we humans *are* animals, although if we admit there is a biological arrow of time toward increasing complexity we can say that we are elevated animals.

From that "elevated" position, we can examine ourselves. When we do, we discover something truly astonishing: We have an internal, private world as well as the external, public, and objective world. Among other things, this internal "private" world produces feelings and emotions in conjunction with our own body and the outside world. We experience the external world and our body always accompanied by our feelings.

We can throw away the feelings, saying that they are subjective and inconsequential, as some materialists do, but we would be too hasty. Upon consultation with other humans, we find there is a spectrum of emotions and feelings that all of us experience. For all of us, in some measure, our feelings affect our actions.

Then we start to wonder. Could animals also have this internal world of feelings? At what stage of evolution do they become a factor in animal behavior? As we don't seem to have any access to the internal world of an animal, how do we investigate the effect of feeling on animal behavior?

Now the materialist biologist can make a valid point. "You are being too hasty. You accused us of robotomorphism. Now we will accuse you of anthropomorphism — the tendency of humans to impose upon the world prejudices that arise from specifically human experiences. How can one do science with such prejudices?"

Is a middle ground possible? The philosopher E. S. Russell (1930) explicates:

> Biology occupies a unique and privileged position among the sciences in that its object, the living organism, is known to us not only objectively through sensory perception, but also in one case directly, as the subject of immediate experience. It is therefore possible, in this special case of one's own personal life, to take an inside view of a living organism. Naturally, the direct intuitive understanding obtained by this immediate experience of living cannot form the subject-matter of biology, which is an objective science, and we must be very chary of reading into other organisms the motives and the modes of experiences which we discover in ourselves. Nevertheless introspective knowledge does give us an insight into the reality of the living organism which cannot be otherwise obtained, and supplies us with a standard by which to test our conceptions of the living thing. (Quoted in Augros and Stanciu 1988, 83–84)

We have come very far since 1930, when these comments were written. As the discussion on chakras shows, our feelings are associated with our internal organs. Logic dictates that any organism having these organs will have similar associated feelings.

In a later section of this chapter, I discuss the possibilities now being investigated for the direct measurement of vital energies associated with our feelings and emotions. In anticipation, I claim the following: The development of a new biology of feeling is now under way.

The Problem of Recognition

Harman and Sahtouris (1998) have identified recognition, the question of how organisms recognize themselves as opposed to others, as a major unsolved problem of materialist biology. To be sure, some progress has been made on the problem. At the cellular level, an alien cell will give itself away if antibodies appear as an immune system reaction to "not me." For the important problem of recognizing the opposite sex in order to mate, it has been suggested that the sensing of attractor scents, called *pheromones*, does the job. But neither of these solutions cover the entire gamut of the recognition problem.

The solution is to see the role of consciousness and feelings in this question. Even the primitive prokaryotic cell has the capacity of self-reference; it "knows" in some primitive way about its separateness vis-à-vis its environment. With the evolution of organs and the feelings of the associated morphogenetic blueprints, the recognition of self and others takes a quantum leap. If vital energy at a chakra is enhanced, then the "other" is friendly; if vital energy at a chakra is depleted, the opposite is true. Because there is no doubt that animals can distinguish friend and foe, there is also no doubt that animals have feeling; they are not robots.

Acupuncture and Homeopathy: Two Impossible Problems for the Materialist

President Richard Nixon's historic trip to China in 1971 resulted in not only a renewal of trade with that country but also an upsurge of Western interest in traditional Chinese medicine, especially acupuncture. Acupuncture is healing through the superficial insertion of sharp needles in parts of the body called acupuncture points. Ever since Nixon's visit, the question of how acupuncture heals has become a major problem for allopathic medicine, which is based entirely on the materialist approach to reality.

Researchers in allopathic medicine look for an explanation of acupuncture along materialist lines of thinking. For example, one theory of how acupuncture relieves pain is that the inserted needles cause much minor, tolerable pain that takes attention away from the major source of pain; that is, they jam the communication channels of the nervous system.

It turns out, however, that traditional Chinese medicine already had the right explanation in terms of the movement of a mysterious energy called *chi*, if we unabashedly identify chi as the movements of the energy of the vital body (Yanchi 1988). In the Chinese view, chi has two modalities, yang and yin. You can see the parallel with quantum thinking here. Yang is the analog of the wave mode of chi, and yin is the analog of the particle mode. Correspondingly, each organ, and even the entire organism, is classified according to the two modes of how chi is processed: the yang mode, which involves possibility and creativity, and the yin mode, which involves manifestation and conditioning.

We now can understand how the healing of pain takes place in acupuncture. The application of the needles to suitable areas of the physical body stimulates the correlated parts of the vital body. Under this stimulation, yang chi, the creative component of chi, flows through vital pathways, called *meridians*, to the vital blueprints of the major organs. This flow increases the general level of yang chi, creative chi, in the vital body, especially in the vital correlates of the brain areas that produce endorphins, the brain's own opiate pain reliever. In other words, the manifestation of the vitality of chi at the vital level in turn manifests brain states with endorphins.

One can verify the correctness of this picture by injecting an acupuncture patient with endorphin blockers, narcotic drugs that block the action of opiates. Indeed, these drugs neutralize the pain-healing effect of an acupuncture treatment (for more details, see Goswami 2004).

If acupuncture is a relatively recent thorn in the side of materialist allopathic medicine, homeopathy (Vithulkas 1980; Ullman 1988) is a much older thorn. The first principle and the major mystery of homeopathy is its "less is more" philosophy. In homeopathy the medicinal substance is diluted with the help of a water-alcohol mixture to such eventual proportions that on the average not even one molecule of the medicinal substance can be said to be present in the concoction that is administered to the patient. If no medicine is given, how does homeopathy heal?

Allopaths attacked the problem by running many clinical tests to directly disprove the efficacy of homeopathy. But tests have now confirmed that homeopathy does work and not just as placebo either. (A placebo treatment commonly means administering sugar pills when the patient believes that he or she is getting real medicine.)

So how does homeopathy work? First we recognize that the medicinal substances in homeopathy are organic, that is, they have not only physical bodies but also vital bodies. The physical body is diluted away (and good riddance, usually, as this part is often poison to the human body), but the vital body is preserved.

How is the vital body preserved in the eventual medicine? To solve this mystery, we need to look at the details of how a homeopathic medicine is prepared. You take one part of the medicinal substance and dilute it with nine parts of the water-alcohol mixture. Then you take one part of the new mixture and dilute it again with nine parts of the water-alcohol mixture. And you repeat that dilution thirty times, one hundred times, even a thousand times to get a homeopathic medicine of increasing potency. The procedure sounds innocuous until we realize that we have missed something. At each stage of dilution, the medicine and the water-alcohol mixture is thoroughly shaken. The word *succussed* is used for all that shaking, and herein lies the mystery. The succussion transfers, through the intention of the preparer, the vital energies of the medicinal substance to the water in the sense that the water of the water-alcohol mixture becomes correlated with the vital energy of the medicinal substance. As a result, whenever we get the home-opathic medicine, although we don't get any of the physical part of the medicine, we do receive the original vital energy with the correlated water. It is also said that the more the homeopathic remedy is diluted, the greater its potency. Why? More dilution means more succussion, which means better nonlocal correlation of the water with the vital energy of the medicinal substance: hence, greater potency.

If the disease is at the vital level, due to vital energy imbal-ances, then homeopathy will work better than allopathy. Home-opathy addresses the vital energy imbalance directly through the application of the vital energy of the remedial medicine.

The choice of the remedial medicine is governed by the sec-ond principle of homeopathy: "like cures like." If the medicinal substance incites symptoms in a healthy body that match the symptoms of the diseased body, then the vital energy movements of the medicinal substance and the body's relevant (unbalanced) vital energy movements must be in "resonance." In that case, the vital energy of the medicinal substance will balance the imbalance of the diseased person's vital energy.

We see that the main principles of homeopathy can be under-stood by using vital body science, and many more details can be explained in the same way (Goswami 2004).

THE PROBLEM OF HETEROGENEITY
AND VITAL TYPOLOGY

Another concern that can be satisfactorily explained by dynamics in the vital energy is the issue of heterogeneity, how individuals can differ so greatly not only in form but also in function. Genetic determinists don't worry about biological heterogeneity because for them DNA programs determine the organism completely. Similar DNA programs should produce similar structure, function, and behavior. The differences in these quantities can be traced down directly to the difference in the genes.

In effect, this is the principle of homogeneity as it holds in physics and chemistry—similar micro ingredients, similar macro effects. Of course, this kind of simplistic expectation based on the physical sciences fails in biology. For example, by this token, the human genome should be very different from the genome of mice, but the opposite is true.

In contrast, organismic scientists such as Walter Elsasser think that biological systems are so complex that heterogeneity of the individual and of classes is one of the factors that restricts physicochemical explanation of biological phenomena. If organismic form (structure and function) is chosen to conform to blueprints of morphogenetic fields, and there is freedom of choice, then indeed we can expect heterogeneity to have the upper hand over homogeneity. In that case, how do we develop and apply general scientific principles in biology and medicine?

Fortunately, for systems with prior experience recorded as memory, as I have already mentioned, the freedom of choice is not always so free. In its total freedom, the application of freedom of choice in the event of quantum collapse is called *fundamental creativity*. When the freedom of choice is restricted to previously discovered contexts or combinations thereof, the event of collapse is one of situational creativity (see chapter 4). If freedom of choice is not exercised in favor of a previously learned response, the event of collapse can be said to be determined or conditioned (using the language of psychology).

In this way, even in heterogeneous biological organisms we can expect a certain regularity in the way forms, functions, and behaviors come about. But determinism (even nongenetic determinism) is too simplistic. Complete conditioning characterizes just one type of biological organism; there are others.

In the Indian system of medicine, ayurveda, individuals are classified according to three *doshas*, or "defects" in Sanskrit. How do the doshas come about? They arise from the way consciousness uses the vital body morphogenetic fields to construct biological form in the physical body. The important point here is that the environment (internal and external) is always changing, and the forms of the physical body are affected by the changes. The morphogenetic fields correlated with the forms adjust to these changes. If the adjustment required is very great, fundamental creativity (when applied to the vital body, it is called *tejas* in Sanskrit) will be necessary. If the adjustment is moderate, only situational creativity (for the vital body this is called *vayu* in Sanskrit) will be needed to bring the morphogenetic fields in line. And if the changes in the physical are so minimal that one can get by with the existing morphogenetic field, then inertia (or the existing conditioning called *ojas* in Sanskrit when applied in the vital context) prevails.

In the formative years of the organism, excess use of fundamental creativity (tejas) for the adaptation of the morphogenetic fields to environmental changes produces the ayurvedic dosha of *pitta*. Excess use of situational creativity (vayu) produces the dosha of *vata*. Finally excess ojas produces the dosha of *kapha*. Ayurvedic physicians make much use of the doshas in their diagnosis and treatment of disease, so their validity is based on extensive empirical data (Lad 1984).

MEASURING VITAL ENERGY

Can we measure vital energy with a physical instrument? The answer is no, by definition. Vital energy and physical instruments belong to two different worlds that do not directly interact. There's a work-around, however. The physical forms in biological organisms represent the vital body morphogenetic fields and are correlated with them. If we can measure these correlated physical forms as they change with the movements of the vital body, then indirectly we are measuring something about the vital body. We already use this strategy to measure thinking. Can we tell if somebody is thinking? Yes. We look at activity in the brain with a magnetic resonance imaging (MRI) device or with positron emission tomography.

I think maybe we already have a method for measuring the vital body indirectly by measuring physical changes: the controversial technique of Kirlian photography. Kirlian photography was discovered by the Russian scientists Semyon and Valentina Kirlian. It involves the use of an electric transformer called a Tesla coil, which is connected to two metal plates. A person's finger is placed between the plates, where it touches a piece of film. When the electricity is turned on, the image the film records is called a Kirlian photograph.

Typically, Kirlian photographs show an "aura" around the object. Proponents of Kirlian photography claim that the color and intensity of the aura are descriptors of the emotional state of the person whose finger is being used in taking the photograph. For example, a red and blotchy aura corresponds to the emotion of anxiety, a glow in the aura indicates relaxation, and so forth.

Clearly, some sort of energetic phenomenon is taking place. It has been verified that the energy involved cannot be controlled by the five senses. Originally, some researchers thought the pictures show subtle energy flow from the finger to the film via psychokinesis. But this could only be true if subtle energy were physical somehow. An alternative materialist explanation has also been given—that the auras are related to sweating. Indeed, the presence of moisture between the plates affects the photographs, creating controversy.

I bring all this up because it is possible to give a third explanation. Changes in vital energy, as in mood swings, do change the programs that run the organ representations, so their functions also change, reflecting the mood swing. The photograph is measuring a change in the physical level, but because the physical-level changes are correlated with the vital level changes, indirectly we are measuring the latter. Indeed, measuring vital energy changes with Kirlian photography has now become so sophisticated that it is being used for diagnostic purposes (see chapter 15).

QUANTUM NONLOCALITY: THE PUBLIC ASPECT OF VITAL ENERGY

Materialists make the basic assumption that feeling is always internal and private. But if feelings are quantum modes of vital energy, then the possibility exists for nonlocal sharing of feelings and vital energy between suitably correlated living beings. If a certain experience is a shared experience, we can no longer say it is internal and private. Because we can verify its existence from a variety of sources and arrive at a consensus, the experienced object must be considered as "external" and "public" in the same sense as a material object is external and public. For this reason, quantum nonlocality experiments involving sharing of vital energy (feeling) between us humans and plants and animals are some of the most convincing evidence for the independent (nonmaterial) existence of vital energy.

The researcher Cleve Backster became a popular icon in the 1970s because of his revolutionary demonstrations of feelings in plants. At the height of the phenomenon, the television gurus Johnny Carson and Merv Griffin invited Backster to be a guest on their shows. As usually befalls a popular icon, he was vehemently attacked as a wacko by other scientists; the establishment biologists claimed they could not replicate his experimental data.

Backster's experimental procedure was very simple. For example, in one experiment, one of his students went into his laboratory and totally destroyed a certain potted plant in the presence of another plant without any human witness in the room. Later,

Backster asked all nine of his students to come one by one to the laboratory, but now he attached polygraph sensors to the surviving plant. As you know, polygraph sensors react to emotional responses in humans. Amazingly, when the plant destroyer entered the room and encountered the surviving plant, the plant reacted, as registered by a large spike in the polygraph chart. If a human were being tested, a polygraph tester would unequivocally declare from the polygraph data that the human was exhibiting fear. Why should the interpretation be any different for the plant?

Two comments: First, plants feel, making it highly plausible that animals also feel, as I have discussed earlier. Second, the surviving plant was not itself attacked. What was the source of its fear of that student, then? The answer is its quantum nonlocal vital body connection with the destroyed plant. The experiment proves feelings are nonlocal, and hence objective, because through nonlocality feelings can be shared by more than one living being.

As to the accusation of nonreplicability, Backster was certainly no lone wolf in the business. He was just following up on work by the biologist Bernard Grad, who did some pioneering experiments on the nonlocal communication between humans and plants and also between humans and animals (Grad 1964, 1965; Grad et al. 1961). Other work has shown similar results. In a series of experiments carried out in China, chi gong masters were asked to throw good chi (associated with good feeling) toward plants. The plants' metabolic growth rate was measured and showed significant improvement. Indeed, when the same masters threw bad chi at the plants, the poor plants suffered deterioration as measured by their reduced metabolic growth rate (Sancier 1991).

The biologist Rupert Sheldrake (1999) has also demonstrated quite conclusively that pet dogs have nonlocal communication with their masters. Sheldrake's motion picture on this work is particularly convincing. At exactly the moment a dog's master gets up from her desk in an office to return home, the dog walks to the window and begins watching for her. How is this possible? The experimental control ensured there could not possibly be any local communication between the dog and its master. There must have been nonlocal communication, most likely via sharing of feeling. Sheldrake has now repeated his experiments with humans and parrots, with similar astounding results demonstrating nonlocal communication.

PLANTS AND ANIMALS
DO FEEL

Mainstream biologists' negative response to experiments such as those described above is selfishly motivated. If animals feel pain, it is hard to justify using them in laboratory experiments that invariably inflict pain. Should we give up on laboratory use of animals, then? We have to be careful about making blanket rules. Certainly, we have learned much from the use of animals in the laboratory. Furthermore, we do kill animals for food, which also inflicts pain. What is the difference?

The difference should be obvious. We eat animals to survive, following a well-known tradition of the animal kingdom in which animals routinely kill other animals for food and survival. So long as we share "animality" with animals, so long as we are motivated by our base emotions of survival, becoming vegetarians en masse is not going to happen (see chapter 21 for further discussion).

But we don't necessarily need animal experiments in the laboratory for survival; some guidelines can be established here. Some spiritual traditions have established rituals used before one partakes of animal flesh. Similar rituals should be established when some survival necessity (for example, fighting an epidemic) dictates the use of laboratory animals.

Finally and fortunately, I believe that the nonbiologist public will be quite happy about these results. In the seventies, it became quite fashionable to hug trees and sing to indoor plants. I hope the reader will consider going back to such rituals, feeling validated by much good theory and good data.

PART 5

NEUROSCIENCE
within
CONSCIOUSNESS

CHAPTER

15

The RELATION of
CONSCIOUSNESS, the BRAIN,
and the MIND

W hy talk about neuroscience in a book about evolution? When we see evolution as changing the representations of the subtle possibilities of consciousness, we can see that the topic goes beyond evolutionary biologists' focus on the evolution of form, that is, evolution of the representations of the vital morphogenetic possibilities. In the new view, evolution is also about the evolution in making representations of mental meaning. Here is where the brain and neuroscience become involved. In this chapter and the next, I will discuss how to study the brain within the primacy of consciousness and how this approach resolves some of the important paradoxes of neuroscience, including the "hard question" of consciousness and the paradox of perception. In chapter 17, I pick up the subject of the evolution of the mind.

Neurophysiologists study the brain and vainly hope that their studies will lead to an understanding of consciousness as a brain

245

epiphenomenon. Alas! Their hopes will never be realized because the interaction of objects, in this case neurons, can only produce an object. The most wonderful aspect of consciousness is not that objects are seen in consciousness but that a subject appears to do the seeing, that there is a subjective experience. Explaining how this subject-object split awareness can arise starting with genes or even neurons is an impossible problem for the materialists. The impossibility mounts when we recognize that some of our subjective experience has causal efficacy. This problem of a causally efficacious consciousness cannot be solved within materialist metaphysics.

We have to venture outside the materialist arena and turn the materialist metaphysics upside down. Instead of saying that consciousness is a brain phenomenon, we have to posit that consciousness is the ground of being, rich with quantum possibilities, of which material possibilities are a component. This underlying reality we cannot experience. Experiences are conscious events involving awareness of the subject-object split; they take place when consciousness chooses an actual event out of these quantum possibilities.

Is there a layer of reality of which we are not conscious, that is unconscious in us? The answer is, of course, yes. The scientist who intuited this other reality and revolutionized psychology with his insight was Sigmund Freud. As we will see, quantum physics can help us understand the mysterious reality of the unconscious.

DISTINGUISHING UNCONSCIOUS AND CONSCIOUS: QUANTUM MEASUREMENT IN THE BRAIN

How do we perceive a stimulus that involves measuring the stimulus? We have to recognize that in every event of perception, that is, at every event of a quantum measurement, we measure both the object we perceive and our own brain. Before we measure, the object is a wave of possibility, so the stimulus the brain receives from the object is a stimulus in possibility. Upon receiving a stimulus in possibility, the brain, too, becomes a wave of possibility, a superposition of macroscopically distinguishable possible brain states. When we choose the state of actuality of the object we perceive, we choose from among the possible brain states as well.

You can see the paradox here. The brain (and the object/stimulus) remains in possibility until a choice among its possible states has been made. But without a manifest brain, there is no observer, no subject's "I" that is doing the choosing. Thus we have a logical circularity, a paradox.

We have encountered such a circularity before, in our discussion of the living cell. We call it a tangled hierarchy. Quantum measurement involving a living cell is special because there is a tangled hierarchy in the measurement process of the living cell. Similarly, quantum measurement in the brain is special because there is a tangled hierarchy in the way the brain participates in a quantum measurement.

In the living cell the tangled hierarchy is most easily seen between the DNA molecule and the proteins. No simple, hierarchical, step-by-step process starting from the microlevel can produce the macromolecules of DNA and protein, which therefore form a tangled hierarchy: It takes DNA to make protein, it takes protein to make DNA. In the brain, the tangled hierarchy involves two similar systems.

Neurophysiologists try in vain to decipher the stages of the processing of a stimulus by the brain. Consider an optical stimulus. A light quantum (photon) from the object arrives at the retina of an eye, then travels a nerve as an electrical stimulus to a brain center, etc., etc., or so the neurophysiologists tell us. To their credit, they can do the analysis for a bit (just as the origin-of-life theorists can do their analysis for the synthesis of a cell for a bit), but then everything gets jumbled up. The brain is too complex.

We can, however, recognize what is involved in a general way. For a quantum measurement to take place in the brain, two things need to happen. First, a series of intermediate apparatuses must process and amplify the stimulus, taking it from the microscopic scale to the macroscopic scale (the perception apparatus); and second, another series of apparatuses must make a macroscopic image, or memory, of the stimulus (the memory apparatus). The tangled hierarchy is involved in the micro-to-macro transition in the creation of both the perception and the memory apparatuses. As a result, we have two apparatuses at the macro scale with a circular relationship: Perception requires memory; memory requires perception. We cannot trace this relationship to the microlevel all the way, however; we cannot reduce it to step-by-step material interactions. These apparatuses have formed a whole that is greater than its parts, and their complexity is of the irreducible type; in short, they are a system of tangled hierarchy.

Finally, what we perceive is the object, that is, the source of the stimulus. We perceive neither the image or memory in the brain nor the brain state that includes the memory; instead, we *identify* with the brain state that has been collapsed in the process of observing and experience ourselves as a subject looking at the object.

Whenever there is such a tangled hierarchy in a quantum measurement situation, there is also self-reference: the codependent appearance of a subject that sees and the object that is seen.

How can all this help us make the distinction between the unconscious and the conscious? An unconscious state occurs when there is processing of possibilities but no collapse. Possibility objects interact with other possibility objects, expanding in overall possibility. There is consciousness and processing but no manifest awareness. I call this a state of unconscious processing—processing of possibilities without collapsing them into a subject-object experience that we call awareness. In contrast, conscious processing involves collapse, which creates the subject-object split that constitutes awareness.

Essentially the same argument holds for a mental object of meaning. We must recognize that the mental object cannot collapse of itself. The arena of the mind doesn't have a division between micro and macro, so no tangled hierarchy can be generated; no transition from the micro to the macro occurs. However, mental meaning can correlate with a physical object. When the physical object is collapsed, the correlated mental meaning must collapse also. Consequently, the brain memory that results from a particular collapse event is not only a memory of the physical object but also of the mental meaning, since evoking the memory simultaneously evokes the correlated mental meaning. In other words, the brain has made a representation of the mental meaning.

Going back to Freud, the eminent scientist did create a little bit of confusion with the wording of the concept of the unconscious. From the discussion above, clearly what he called "unconscious" he should have called "unaware." Since consciousness is the primary reality, it is omnipresent; it is the ground of all being, so where else would it go? What comes and goes, what appears discontinuously in the event of quantum collapse, is the experience of a subject-object split that we call awareness.

One word of caution, however. Freud's concept of the unconscious is actually much narrower than that indicated by quantum physics. To see why, consider an object (mental, physical, or both) that we have never before experienced, an unlearned stimulus, so to speak. When quantum consciousness collapses this kind of original stimulus, the object is experienced in its suchness, in a consciousness that is cosmic and creative, unconditioned. The unconscious processing that takes place before such a collapse event is likewise free, unrestricted, unlimited by conditioning. This view of the unconscious is akin to the concept of collective unconscious introduced by Carl Jung (1971). In contrast, what Freud originally meant by unconscious can be called our personal unconscious, because its possibilities are built out of personal memory.

I will mention something else in passing. Memories accumulate in the brain as we learn our stimuli. More and more, our
unconscious processing consists of processing our memory. The
tendency is to process every stimulus by reflection in the mirror of
memory. Soon we develop a habitual pattern in the way we use
our mind to give meaning to our experiences, a pattern that we call
our character. This character, plus the accumulation of our memory or personal history, is the quantum physics version of what
psychologists call the personal self, or the ego.

THE BRAIN, GENETIC DETERMINISM, AND DARWINISM

Genetic determinists would have us believe that all program-like
behavior at the macrolevel is the result of genetic programs,
which themselves are the result of Darwinian evolution through
chance and necessity. How credible is this assertion?

Can genetic programs alone account for the biological structure and function of the brain? Arguments akin to those in
earlier chapters can be readily given for any brain function to
demonstrate the small likelihood that such a function would be
the result of genetic programs and chance-and-necessity evolution. In other words, genetic determinism and Darwinian evolution are not a credible explanation of brain functioning. Brain
functions are program-like, no doubt, but this behavior is not
the result of programs at the genetic level whose job ends in the
construction of the building blocks—the essential proteins.

There is another angle to consider. Ask yourself whether the
capacity of the brain to make representations of mental meaning
could originate in the genetic programs. In effect, the materialists
are asking you to suspend your disbelief twice: first, at the level of
biological function at the macrolevel of the brain (for example,
the brain's ability to make memories); and second, at the level of

using memories to represent mental meaning. This second capacity of the brain is so much like a computer that no neuroscientist would deny it. And yet, the proponents of genetic determinism and Darwinism are asking you to believe that the program-like behavior of the brain is also teleonomy, only appearance. Underneath, the genes are responsible for it all.

A further point against genetic determinism comes from the tangled hierarchy required for the observed self-reference that arises in connection with the brain. Self-reference requires tangled hierarchical interactions of components brought together discontinuously, something no genetic program, being a continuous entity, can ever accomplish.

MEANING AND THE RELATION OF MIND AND BRAIN

The neocortical part of the brain involved with mental phenomena such as thought is a computer of sorts. So, materialists ask, can we build a computer with a mind? If we can, that would prove that our mind belongs to the brain. Thus originated an entire field of study called *artificial intelligence.* The mathematician Alan Turing claimed that if a computer can simulate a conversation intelligent enough to fool a human being into thinking that he or she is talking to another human being, then we cannot deny the computer mental intelligence.

So have computers passed the Turing test? A computer has defeated one of the world's greatest chess players, so maybe the computer is even more intelligent than the human being? Not only do we seem to have built a computer with a mind, we seem to have built a computer with a mind better able to reason than one of our best.

The philosopher John Searle (1994) pointed out that a computer, being a symbol-processing machine, cannot process meaning. You can reserve some of the symbols to denote meaning—call them "meaning symbols." Then you need other symbols to tell you the meaning of the meaning symbols, and so on, ad infinitum. To process meaning, you need an infinite number of symbols and machines to process them. An impossible task! In a similar vein, the physicist and mathematician Roger Penrose (1989) gave a mathematical proof that computers cannot process meaning. Giving his book the provocative title *The Emperor's New Mind*, Penrose warned us that, all the hoopla notwithstanding, the computer's proposed new mind is as false as the emperor's new clothes in the famous fable. For his proof, Penrose used Gödel's theorem—that an elaborate mathematical system is either inconsistent or incomplete—an important theorem that is fundamental in recognizing the cogency of tangled hierarchy (Hofstadter 1980).

Materialist biologists claim that meaning may be an evolutionarily adaptive quality of matter. Searle and Penrose's work convincingly exposes the vacuous nature of such a claim. If matter cannot even process meaning, how can matter ever present a meaning-processing capacity for nature to select, survival benefit or not?

The lesson from all this is that although the mind is clearly associated with the brain, it does not belong to the brain; it is not a brain epiphenomenon. Instead, it is a body independent of the brain, being the meaning-giver of our experiences. Neuroscientists have to come to terms with that reality.

But if brain and mind are totally different, brain being matter substance and mind being meaning substance, how do the two interact? Here we face the old question of dualism again. Mind and brain need a mediator. Enter consciousness and psychophysical parallelism! Consciousness mediates the interaction between mind and brain. They both consist of quantum possibilities of consciousness, mind being meaning-possibility and brain being matter-possibility. Quantum consciousness collapses the possibility waves of both brain and mind to create an experience of mental meaning and at the same time create a brain memory of that meaning, a representation.

You can argue that all this is theory. Where is an experiment? We do have a negative experimental test here: If this theory is correct, then it is impossible to build a computer that can process meaning. So far no computer scientist has been able to build a meaning-processing computer to refute the theory. In other words, the theory is passing the test.

The nature of brain memory, judging from its replay, is a dead giveaway for mind being an entity different from the brain. Evidence for the difference comes from the work of neurophysiologist Wilder Penfield (1976). Working with epileptic patients, Penfield stimulated the patients' memory "engrams" (his term) with electrodes and observed that such stimulation produced not a static memory image but an entire stream of mental memory. The mental memory does not reside in the brain. Instead, the stimulation of the brain memory triggers the correlated mind to play its correlated meaning.

Actually, we can find much positive evidence in favor of the causal practicality of meaning processing, which is the job of the mind. I have mentioned two such practical matters already: creativity (see chapter 4) and synchronicity (see chapter 11). Here I will mention two more: dreams and cancer.

The neurophysiological explanation of dreams—that they are the result of attaching perceptual images to brain white noise (Hobson 1990)—is only the beginning of an explanation. The complete explanation is that mind puts meaning into the "Rorschach inkblot" of brain white noise, sometimes creating quite striking audiovisuals. Dreams are the ongoing story of our meaning life, how meaning unfolds in our lives (Goswami and Simpkinson 1999; Goswami 2008). This connection to meaning explains why dream analysis in the Jungian style, where you assume that every dream character in your dream has the meaning that you give to that dream image, is so useful in psychotherapy.

A dream can heal you when you work with it and appreciate its meaning.

While some dreams offer personal healing, others are creative dreams that can disturb the universe. Niels Bohr dreamed of the discrete orbits of the atomic electrons. The inventor of the sewing machine, Elias Howe, got his crucial idea from a dream in which he was captured by savages and was told by their leader to finish his machine or else. Somehow Howe noticed that his captors were carrying spears which had holes near the sharp end. Upon waking Howe at once realized that the key to his machine was to use a needle with a hole near the point. A third example should be dear to the heart of every neuroscientist. It concerns pharmacologist Otto Loewi's discovery of the experimental demonstration of the chemical mediation of nerve impulses. Loewi got the idea through not one but two dreams. The first time he dreamed the idea and wrote it down in the middle of the night, he could not decipher his own handwriting in the morning. Fortunately, his intention brought him the dream the next night as well. This time Loewi was careful to write down the details very legibly.

For a second example of the causal practicality of meaning processing, consider the important field of mind-body disease (Pelletier 1992). Wrongness in meaning processing can give us serious disease (Dossey 1992; Goswami 2004). I will give you one example of this—how the suppression of emotion at the heart chakra can give us cancer.

Cancers can result from immune system malfunction. We always have some cells in our body that are malfunctioning and dividing uncontrollably. When the immune system is healthy, the thymus gland makes sure that these abnormal cells are regularly killed off.

In the West, most people, especially males, are culturally conditioned to suppress emotions. For example, a man may find it disadvantageous to have his heart chakra open in the presence of a woman he likes because an open heart makes him vulnerable.

She might ask for a BMW, and he might agree! Some other horrible thing might happen! Naturally he picks up the habit of suppressing vital energy at the heart, causing an energy block. A prolonged block at the heart leads to the suppression of immune system activity of the thymus, which can in turn lead to the suppression of the body's ability to kill off abnormally growing cells, and such cells then become cancerous. Indeed, certain types of cancer have been connected with the emotional habit of suppressing the energy of love at the heart chakra. Importantly, very good evidence now exists that when the emotions clear up via a quantum leap in mental meaning that unblocks the vital energy at the appropriate chakra, patients undergo spontaneous healing. They take a quantum leap from disease to healing by using their creative choice (Chopra 1990; Weil 1995; Goswami 2004). Slowly such ideas are making their way into the practice of medicine. For example, pioneering work in India by the researcher Ramesh Chowhan has shown that the vital energy block in the heart chakra can be measured via Kirlian photography (see chapter 14), making it possible to use such measurements for early diagnosis of cancer.

What can we learn from these two examples? If unbalanced processing of meaning can produce a serious disease such as cancer, and if right meaning restores health, we had better take meaning seriously. It is not a mere epiphenomenon!

MARCEL'S EXPERIMENT

I want to end this chapter by describing an experiment that illustrates both the quantum possibility wave aspect of meaning processing by the mind and the efficacy of the quantum collapse model to distinguish between the unconscious and the conscious.

The original purpose of the psychologist Tony Marcel's experiment (1980) was to use polysemous words (word with ambiguous meaning) in a series of three words to check the congruence or incongruence of meaning across the series. He measured the subject's reaction time as an indicator of the relationship between the words in such strings as *hand-palm-wrist* (congruent), *tree-palm-wrist* (incongruent), *clock-palm-wrist* (unbiased), and *clock-ball-wrist* (unassociated). His subjects watched a screen on which the three words of the series were flashed one by one with an interval of either 600 milliseconds or 1.5 seconds between them. For example, flashing the word *hand* before the word *palm* should bias the subject to perceive the hand-related meaning of *palm*, which should in turn improve the subject's reaction time to recognize *wrist*. This effect is called *congruence*. In the incongruent string *tree-palm-wrist*, the biasing word *tree* should trigger the tree-related meaning of the word *palm*, thus leading to an increased reaction time for recognizing the third word, *wrist*. And indeed, Marcel found exactly this result.

However, when the middle word was masked by a pattern, subjects showed no appreciable difference in the reaction times between the congruent and incongruent cases. It was already known that pattern masking prevents conscious processing but allows unconscious processing. What the experiment proves is that in unconscious processing, both of the possible meanings of the ambiguous word palm remain accessible; that is, there is no collapse and no choice and, therefore, no biasing of any meaning. The difference between the congruent and incongruent situations disappears.

Elsewhere I have analyzed this experiment in more detail to rule out alternative models of explanation (McCarthy and Goswami 1993; Goswami 1993). The conclusion is unambiguous: The quantum model for the distinction of the unconscious and the conscious is the right model.

Neuroscience within Consciousness

Neuroscientists spend a huge amount of research effort looking for the elusive "seat" of consciousness in the brain and also looking for the mind in the processing of the brain. Such efforts don't lead to anything concrete except redefinitions and vain claims. I hope that the preceding pages demonstrate that the time has come for a turnabout in the neuroscientists' thinking, in which the brain is regarded as secondary to consciousness and the mind.

The integrative value of such a paradigm shift would be enormous. Right now, neuroscience contributes only to the cognitive-behavioral branch of psychology. In the new view, the enormous progress made in consciousness research via depth psychology (in the traditions of Freud and Jung) and via transpersonal psychology (following Maslow and Assagioli) would come under the experimental purview of neuroscience, opening it to an enormous amount of new investigations. Some investigations in this direction have already begun, for example, research on meditation using brain waves.

Neurophysiologists earlier had discovered that our brain waves show specific signatures of our three major states of consciousness: waking, dreaming, and deep sleep. In materialist neurophysiology, this fact was already hard to explain because no distinction is possible between the conscious and the unconscious. Then came the discovery that the brain wave signature of meditative states is also quite unique, quite distinguishable from the three common states of consciousness (Wallace and Benson 1972). This discovery raised the question, In addition to meditation, are there other distinct states of consciousness aside from the usual three?

In spiritual traditions, there is mention of "higher" states of consciousness called *samadhi* in Sanskrit. In the Hindu tradition, it is said there are two kinds of samadhi. A transient state of experiencing the "oneness of everything" is called *savikalpa samadhi*,

samadhi with subject-object split. In Japanese, this state, a state of subject-object split experience, is called *satori*. More recently, Abraham Maslow called this state the "peak experience." Because of its highly transient nature, measuring a brain wave signature for it may be quite challenging.

However, Hindus also talk about a second kind of samadhi, called *nirvikalpa samadhi*, that is, samadhi without subject-object split. This state is therefore more akin to sleep, which also lacks the subject-object split. I call this state "creative sleep" because in this state consciousness unawarely processes new possibilities, not the old ones of memory processed in regular sleep (Goswami 2008). I suspect that there may very well be a specific neurophysiological signature for creative sleep that can empirically differentiate it from regular deep sleep.

The experimental investigation of the higher states of consciousness will open neurophysiology to a broader worldview. The beauty of the new view is, of course, that it does not leave out anything, not even cognitive science. In the next chapter, we will show how the new view solves the problem of perception, a persistent paradox for the materialist view.

16

RECONCILING COGNITIVE SCIENCE *with the* PARADOX *of* PERCEPTION

Just as biologists cannot define life and cannot explain life's origin, the cognitive scientists who dominate academic research in psychology cannot explain ordinary perception. What is perception? Who perceives? What do we perceive? Can we perceive objects in their suchness? Whence come the subjective qualities (technically called *qualia*) of our perception?

So far our introduction to the new science of consciousness has been dominated by the explanation of "Who perceives?" Indeed, this is *the* hard question (Chalmers 1995) for cognitivists and neurophysiologists, who avoid it like the plague. The other questions of perception raised above are equally difficult for the materialist scientists to address properly. The theorists of this field are forever bogged down in controversies they cannot solve. I will demonstrate in this chapter that the new paradigm of consciousness is as efficient in resolving these controversies about perception as it has been with the question of the subject or the self of perception.

CHAPTER SIXTEEN

COGNITIVE MODELS OF PERCEPTION: DIRECT REALISM AND REPRESENTATION THEORY

Almost every experimental neurophysiologist or cognitivist has the same underlying model of perception, often called the *representation theory*. An object represents a stimulus field that presents our perceptual apparatus—the brain—with a stimulus. The brain processes this stimulus, first with the eye and the retina, then with its higher centers. Eventually, an integrated representation of the stimulus/object is made, defining an image in a field of perception. It is this image that we see.

Many questions thwart the validity of this very reasonable picture. Say you are looking at a big cat. You see a big cat, no doubt. However, your brain obviously does not have enough room for a direct, life-size representation or image of this "big" cat. Where is this image that you see located? Furthermore, the representation must be made of neuronal activity of some sort. How do your neuronal activities add up to a big cat that you actually see? Also, who and where are you? Are you assuming there is a TV-like screen with a real picture on it (of the external object) in the back of your head and somehow you are looking at it? But if you think of yourself as a homunculus (a little replica of yourself) doing the looking inside the brain, either you are succumbing to dualism—you separate from the brain—or you get into an infinite regression—who is looking at the homunculus? Ad infinitum.

Experts called identity theorists tackle the questions "what do you see?" and "who are you that sees?" simultaneously by positing squarely that in every event of perception you do really see part of your brain. You and what you see all come out of the neuronal activity of that part of the brain. The experience you have and your brain's neuronal activity are identical. But is this assertion credible? As one theorist (Smythies 1994) puts it, "How many identity theorists really believe it or apply it to their own daily lives?"

There is, after all, another way to get out of the infinite regression (Gregory 1981). Clearly, receiving information from an exterior environment requires making an image. But why should we assume that retrieving information from an image (an internal object) requires making an image, too?

There is still another serious problem with the representation theory. Sophisticated techniques for brain imaging are showing that the perception of an external object often consists of integrating the neuronal activities of many widely separated brain areas. How does the brain bind them together to give us an integral whole that we experience? How do we explain the unity of experience? This is called the binding problem.

There are four other, more philosophical objections that can be raised against the representational theory (Smythies 1994) and in favor of a model of "direct" perception; I list them here:

1. If all we experience is sensations in our brain — inside stuff—then why assume outside physical objects at all? Why not say that there is nothing but me and my sensations? Why not succumb to solipsism, or at least the kind of (dualistic) idealism posited by Berkeley (an eighteenth-century philosopher and Anglican bishop)?

   This latter idea requires some clarification. At a time when Descartes' version of mind-body, internal-external dualism dominated Western thinking, Bishop Berkeley revived the philosophy called *idealism*. Berkeley made the point that we receive all information about the so-called external world through our sense experiences, which are internal. Because we have no way to directly verify the reality of the external world of matter, why postulate it at all? Why not assert that of the mind-matter duo, only mind is real?

   Berkeley's philosophy is well known in connection with the falling tree puzzle: If a tree falls in a forest but no one is there to hear the sound of the tree falling, is there a sound or not? Newtonian physics says there must be a sound, whereas Berkeley seems to be saying there is not a sound because there is no "mind" around. The puzzle was created to discredit Berkeley's idealist philosophy in favor of (material) realism. Incidentally, Berkeley did have an answer to the puzzle: There is a sound, because God's mind is always present, even in the forest with no one there! But that answer only showed the true color of his thinking, his dualist prejudice. You see the problem: It retains the God-world dualism while unifying the mind-matter dualism, and the usual arguments of material monists against dualism can be raised to refute Berkeley—how does God interact with the world, etc.

2. How can we tell that the neuronal activities of my brain (the representation) really represent an external object if we can never directly see and compare them with the object in its suchness?

3. If the objections above seem lethal enough, why not posit that we (that is, the brain) directly perceive the object. This choice gives us the philosophy of direct realism: External objects are real and the brain directly perceives them without the intermediary of some internal images of them.

4. All our knowledge—about brain and about perception—comes from perception. Is it fair to use knowledge obtained by perception to refute the model above of direct perception? That's like cutting off the branch we're sitting on.

Does direct perception make more sense than the representation model? Well, even the direct perception model does not explain how subjective experience can arise from an object interacting with an object. There is the old "hard question" again.

There are other serious problems with the direct perception model as well. Some cases clearly involve the properties of the representation-making capacity of the brain. For example, in color perception, is the color a property of the object? Most researchers now think color is a property of both the object and the brain representation. Moreover, very importantly, direct perception does not explain the subjective qualia—a technical term philosophers use to connote the specific quality of a subjectively felt, first-person experience—of our perception.

We are forced to some inescapable conclusions, which is why people incessantly debate about how we can perceive at all. That the brain makes representations and the representations have an effect on what we see is undoubtedly true. On the other hand, it's hard to rationalize an explanation based on representation that relies on a "television screen" on which a homunculus in the brain views pictures that resemble the outside object. The binding problem is a hard problem for the representation theorist as well. Finally, the philosophical problems of solipsism and dualistic idealism, and additionally, the problem of comparing with direct perception all loom large.

Ultimately, the philosophical debate is one between (dualistic) idealism and (direct) realism. Dualistic idealists see the world as primarily ideas. To them, perception happens primarily because of what we see inside of us. Idealists are the pure brand of representation theorists. Realists support direct realism of perception of objects that are outside and real. They want to avoid any reference to objects that are internal, such as the representations of objects inside the brain.

Cognitive scientists and neurophysiologists are often caught straddling this fence because with the advance of science they are in a position to study the former black box of the brain, and these studies are giving them (and us) obviously interesting results. But they cannot solve the conceptual quandary of which "ism" is right—(dualistic) idealism or realism? Philosophically, most cognitive scientists support some form of realism. And yet, without some emphasis on "internal" representations, how can they justify their trade?

Clearly, pure idealism is susceptible to solipsism and pure realism is impractical. So from this view alone, it is desirable that we have a middle ground established giving some validity to both. Below I will show that the philosophy of monistic, not dualistic, idealism can incorporate realism in such a way as to offer a solution to the philosophical problems of perception.

Do We Have a Big Head?

The hint for how to accomplish such a rapprochement came from two philosophers from different eras, Gottfried Leibniz in the seventeenth century and Bertrand Russell in the twentieth century, who asked themselves the question, Can realism and idealism both be valid in how we see things? Their answer, though, was partially playful; it could not be taken seriously at the times they proposed it.

The idea is quite simple. Realism says that only the external object is real; only objects that we find "outside" of us are real because they are public and we can get consensus about them and make them the object of objective scientific scrutiny. Idealism says that we cannot directly see what is "outside" without the help of the intermediaries of our "inside" private representations. So these inside representations must be more real than the objects they represent. Or rather, they had better be real, because objects in their suchness we will never know.

Easy solution, said Leibniz and Russell (for a good discussion, see Robinson 1984). Suppose we have a "big" head in addition to the small head that we normally experience, so that so-called outside objects are outside the small head but inside the big head. Then aren't both realism and idealism valid? Realism works because the objects are outside (the small head); idealism works because objects are also inside (the big head).

Sounds like sophistry, doesn't it? It does until you process the solution through quantum thinking.

Quantum Consciousness and a Workable Model of Perception

You can recognize our "big" head in quantum terminology as our capacity for nonlocal processing that includes all "small" heads. In other words, when we choose in quantum consciousness or God-consciousness, we are operating from the big head and all objects are "inside" us. The choice collapses the wave of possibility of an object and also the wave of possibility of our small head, the brain. We identify with the brain-state thus collapsed and do not see the brain state as an object. We only see the object/stimulus and see it separate from us, giving us a "spiritual" experience of immediacy in which the object is seen in its suchness.

However, if the object/stimulus is one previously experienced, we do not usually recognize this "primary" collapse event. Instead, we see the object upon repeated reflection from the mirror of its past memory, which is subjective and individual. The past memories modulate these secondary collapse events so that the perception of the collapsed object acquires an individual flavor. This process gives us the subjective qualia of perception.

Experiments show that the processing time of secondary collapse events is 500 milliseconds or thereabouts (Libet et al. 1979). When we finally recognize the object, we are quite identified with our past memories that we have just sifted through; we are operating from the ego-self conditioned by these memories, from the "small" head of individuality. In this perspective, the material object is seen outside us, because of the apparent fixity of the macrophysical world, as explained earlier.

What about that television-image-in-an-inner-theater aspect of the representation theory? We have been forgetting something so far. Along with the external physical object and the observer's brain, there is something else that quantum consciousness collapses routinely—the mind that gives meaning to the observation. The brain representations of the stimulus are literally brain neuronal states, no doubt. They are not unlike the electronic movement on a TV screen. But we do not see the electronic patterns when we watch TV, do we? Instead, consciousness uses our mind to give meaning to the pattern of fluorescent spots on the screen produced by the electronic movement. Similarly, in the case of perception, consciousness uses our mind to give the meaning to the neuronal configuration of each brain representation. Eventually, it is our consciousness, with the help of the mind, that produces the recognizable image of the object of primary perception from the current neuronal representation as modulated by the previous memory.

The oneness of experience of the binding problem is also solved in the model presented here. Because consciousness is fundamentally nonlocal, it can bind together all the different brain areas to produce one unified brain state of collapse.

So the final solution based on quantum physics and monistic idealism as presented here combines the best of realism, (dualistic) idealism, direct realism, and representation theories. At the same time, it explains the suchness experiences that are denied in the Western philosophical tradition but that have long been recognized as spiritual experiences in all the other major traditions. The theory also explains qualia of normal perception and solves the binding problem. And, of course, the problems of the subject-object split in awareness and the ego-modality of normal perception are also explained from the outset.

Now do you see why, without God, we couldn't see anything? Obviously that is true, not just about seeing, but about any experiences with our senses. Now you can appreciate the following quote from one of the Upanishads (Mahadevon 2000):

> That immutable . . . is the unseen seer, the unheard hearer, the unthought thinker, the ununderstood understander. Other than it there is nothing that sees. Other than it there is nothing that thinks. Other than it there is nothing that understands. (154)

A BRAIN IN A VAT

The neurophysiologist Jonathon Harrison (1996) has given us the interesting tale of Ludwig—a brain in a vat—to ponder. Ludwig, who was born with a great brain but a deformed body, was "rescued" by a surgeon, Dr. Smythson. This surgeon removed Ludwig's brain from his ugly body and kept it functioning with a life-support system. The good doctor didn't stop there. The cut ends of Ludwig's cranial nerves, spinal chord, and all that were connected to a complex cybernetic system of such capacity that Ludwig's brain could continue to be stimulated in exactly the way it was before via sense organs and external stimuli. Dr. Smythson did not forget to connect the cut ends of the motor nerves and the "sensory" input so that Ludwig also experienced the so-called willed movements. Thus Ludwig, though he was a brain in a vat, so to speak, was able to have all the experiences of a normal human life—eating a meal, sleeping, thinking philosophy, having a love life—whatever the good surgeon cared to simulate, including all the usual experiences of a normal life with a real physical body.

So Harrison's question is, How do we know that we are not brains in a vat? It seems that we don't know, and we can't know!

You can see that this is the same argument that seems to justify solipsism or Berkeley's kind of idealism. The truth is, all local experience can be simulated. Today, we can readily imagine constructing a computer-generated virtual reality that can replace our local reality.

So what is the way out of this kind of paradox, if it is a paradox? For a monistic idealist working from the theory of perception we have constructed above, there is no paradox. Computers can simulate everything that is local, but nothing that is nonlocal. So Ludwig's so-called simulated life would entirely consist of local experiences. The nonlocal mixture that gives us the spark of life, like the oceanic feelings of love, would be completely lacking.

Think again of that previously mentioned *Star Trek* story. What man would prefer many beautiful android women to one human woman?

COGNITIVE SCIENCE WITHIN CONSCIOUSNESS: FURTHER COMMENTS

Cognitive scientists want to make mind science more objective, a goal toward which they have made considerable progress through, for example, magnetic resonance imaging of thoughts. We have come a full circle. The lesson of all this is that we cannot reconcile perception itself with cognitive science without firmly founding perception within the metaphysics of the primacy of consciousness, without giving value to what is internal.

It is time to cease the battle for supremacy between external and internal and recognize that science needs both. An MRI image of thought does not give the scientist any clue about the meaning content of thought being observed; for that, an intersubjective dialog between the scientist and the subject of the thought is essential.

Will cognitive science still be science if subjective experiences of meaning are allowed to enter it as valid data? Yes, so long as the criterion of strong objectivity is replaced by that of weak objectivity. Strong objectivity demands that data be independent of consciousness, of observers. This position is not tenable. Weak objectivity demands only observer invariance, that data be similar from one observer to another. The cognitive experiment of Marcel (1980) described in the last chapter is a good example.

17

The EVOLUTION
of the MIND

*I*f the mind were an epiphenomenon of Darwinian evolution and of the brain, then any changes in the way the mind operates could occur only continuously, from continuous changes in the brain. Also such changes could not be progressive; they could only be haphazard. Empirical observations tell a different story. In the evolutionary history of the mind, there have been three distinct, and progressive, stages.

EVOLUTION OF THE VITAL AND MENTAL REPRESENTATIONS

Materialists talk about the stages of evolution of the biological form as a progression from one-celled organisms, to multiple-celled organisms, to invertebrates with well-defined organs, to invertebrates with a nervous system and sense organs, to vertebrates with a hindbrain, to vertebrates with a limbic brain (midbrain), to vertebrates with a neocortex, to human beings with an advanced neocortex. The philosopher Ken Wilber has pointed out that these stages correspond to the evolution of exterior aspects of consciousness. To see the evolution of the representation of the vital and the mental bodies, we have to look at the interior experience available at each of these stages (Wilber 1996).

What interior experience is available for the lowly prokaryotes? As the biologist Bruce Lipton (2005) has shown, even the prokaryote has programmability in the protein structure of its cell wall (see chapter 8). Accordingly, consciousness programs it for the basic biological functions, giving it the experiences of rudimentary sensations and feelings. According to Lipton, the programmable proteins move to the interior of the cell in the eukaryotes, which have even more programmable organelles. Thus eukaryotes should be able to experience sensations; they can also experience feelings corresponding to the basic biological functions of metabolism (specifically, anabolism and catabolism, giving it the feelings of existence and security) and sexual reproduction (giving it the feelings of sexuality).

With the development of organs, the chakras develop, and invertebrate organisms can experience feelings at the chakras. With the development of the nervous system, perception becomes possible, followed by impulsive movement. With the development of the most primitive brain, the reptilian complex, mentation begins, feelings are given meaning, and vice versa. The effects of feelings on the mind, the emotions, are collectively mapped in the limbic brain as instincts. At this point the organism develops the capacity of mental images and is able to react emotionally to mental images, including archetypes.

With the development of the neocortex, symbol processing begins. Finally, with the appearance of humans and their complex neocortex, the ability to process concepts via the making of mental representations of the supramental archetypes manifests.

When we combine the exterior with the interior evolutionary stages, we get the correspondences listed in table 2.

Table 2. EXTERIOR AND INTERIOR STAGES OF EVOLUTION

Exterior	Interior
One-celled organisms (prokaryotes)	Rudimentary sensations Rudimentary feelings
One-celled organisms (eukaryotes)	Sensations Rudimentary feelings
Multicellular organisms (such as plants)	More developed feelings of existence, sexuality, and security
Invertebrates with organs but no nervous systems	Feelings at well-defined chakras
Organisms with nervous systems	Perception, impulses
Reptiles (hindbrain)	Emotions
Mammals (limbic brain)	Emotions connected with images
Primates (neocortex)	Symbols
Humans (complex neocortex)	Concepts

STAGES OF THE EVOLUTION OF THE HUMAN MIND

In our quantum model, what are the stages in the evolution of the representation-making capacity of the human mind, both in individuals and in the species? Initially, the human mind must be busy making sense of the physical environment—the most stable and compulsory aspect of its experience. Next must come a preoccupation with the vital-body feelings, that is, with giving meaning to feelings. Only then can we expect the evolution of a mind whose preoccupation is mind itself, the meaning of meaning. And finally, the mind can evolve enough to turn its attention to the supramental.

Below I give details of this evolutionary vision, which not surprisingly is close to the philosophical musings of Sri Aurobindo.

The Physical Mind. The mind puts meaning to what the brain takes in as physical stimuli. The mapping of the mental meanings produces developmental software in the brain that we can call the physical mind. At the earliest stages of development, the physical mind is a dominating factor for everyone. But for some people, even after the development of the adult ego, the dominance of the physical mind continues. It is operationally possible to get through the affairs of the world with the physical mind alone, but such a person does not access the rich, inner life available through the other domains of the psyche. For example, a person with such a mind is capable of building a logical structure, but only when he or she can model it from the sensory world of physical objects. Whatever cannot be thus modeled is considered magic.

Someone with this kind of mind is resistant to feelings, too, because he or she cannot make a physical model of the feelings. Women in the traditional role of child rearing have a clear disadvantage if they operate at this stage, because they are forced to attend to the children's feelings.

You can see where this stage fits in the evolutionary history of humankind. Anthropologists identify the early period of human evolution as the hunter-gatherer stage; the males were the hunters and women, the gatherers. Both occupations require attention to the external physical world; hence both men and women of this age acquired a predominantly physical mind. However, the preoccupation of pregnant women and mothers with emotion relegated women to second-class status, and hence patriarchy developed. The males were the interpreters of the archetypal domain; therefore, only male gods were worshipped in this era. Anthropologists also have designated the worldview of people of this era as magical, which is clearly a correct choice.

The Vital Mind. The vital level of mind is the first level at which the extraphysical subtleties of consciousness are recognized. The vital mind consists of mental representations of meanings of vital stimuli that elsewhere I have called the mentalization of feelings (Goswami 2004). Such a mind cannot yet engage the intuitive facility synchronously with the mind to sort out feelings according to the archetypal context in which the feelings arise. As a result, such a mind is lost in the melodrama of instinctual emotions, most of which are experienced negatively. A vital-level mind thrives on sensuality, pleasure, and even pain. Characteristically, the vital mind looks at the world as a dichotomy of likes and dislikes. Among other things, this distinction is a precursor to the good-evil dichotomy of the mind that comes later. However, no contradiction exists between the ways of the physical mind and the vital mind. For a vital mind, the physical mind continues to be useful and is therefore integrated into life.

The vital mind was clearly the way of the human mind in what anthropologists call the horticultural era—a farming culture based on simple machinery such as the hoe. This is one era when both men and women worked at the same occupation. Because feelings dominated both men and women but women, through motherhood, had a clear lead in terms of familiarity with feeling, this was the only age (before the present era) when women had the same social status as men and when matriarchy was perhaps as common as patriarchy. Anthropologists tell us that whereas the previous era had only male gods, in this era one finds both male and female gods being worshipped. Clearly, women were allowed to interpret the archetypes and their interpretation was valued. I think it is quite appropriate that this era is regarded as the golden age by ecofeminists.

Also, the worldview here changed from magical to what is called mythical, being centered around myths of gods and goddesses. The gods and goddesses represented the archetypes of emotions, both negative and positive.

The Rational Mind. When we learn to use the mind at purely the mental level without reference to the physical or the vital, when we learn to analyze the meaning of meaning, we are in a higher mind that I call the rational mind (Aurobindo's term for this stage is "higher mind"). At this stage of the rational mind, we are capable of abstract thinking and reasoning.

The logical structures of the rational mind do not have to depend on mechanical models, although they can. A person of rational mind is capable of looking at the meaning structure of his or her constructs and of taking an occasional quantum leap of creativity to discover a new meaning. The rational mind is clearly hierarchical, and the head of the hierarchy is the personal ego that the person uses to sort out his or her experiences.

The biblical story of the "Fall" is about the evolution from the vital mind to the rational mind. In the vital mind, both Adam and Eve are in paradise. With hardly any ego development, they are one with nonlocal consciousness—they walk with God and talk with God. But all that changes when they eat the fruit of the tree of knowledge and rational mind begins. If you read the actual story carefully, you will find that it is Adam who is banished from paradise, who made the move into rational mind, not Eve. This detail is important because women were denied meaning processing in the era of the rational mind until very recently.

For individuals who live from the rational mind, development takes an interesting path. As children, people belonging to this category typically take many creative leaps of discovery into the supramental (fundamental creativity) to discover the contexts of (abstract) thinking. But for the adult ego, a homeostasis sets in, and this type of mind identity becomes complacently comfortable with the learned repertoire of logical structures or belief systems. At best such people remain capable of situational creativity.

Consequently, a rational mind displays a tendency to simplify and marginalize the supramental origins of its learned repertoire of contexts, conceiving of the world in terms of simple dichotomies—good and evil, beautiful and ugly, true and false, love and hate.

Feelings are not rational, and thus the vital mind is not compatible with the rational mind. Because of this incompatibility, plus their tendency to complacency, people of the rational mind are prevented from integrating the vital mind, which is now seen as a nuisance. Women who are mothers are once again at a clear disadvantage if they operate from this mind because of the importance of emotion in their children's lives.

According to anthropologists, the rational mind shows up in the men of the agrarian era in cultures that depend upon heavy machinery, such as ploughs, that only men can operate. Here men become the producers, and once again women are relegated to the status of reproducers. As such, they are not allowed the privilege of meaning processing. Even among men, only the dominant men of society get to process meaning in a large way. This era saw the appearance of the great kings and warriors and a clearly class-based society.

Notice another important feature of such societies. The excess production of the agricultural era now allowed the existence of a new class of creative people—the Brahmins, to borrow a Hindu term, though the concept applies beyond Hindu culture. Unencumbered by the requirements of making a living, these creative people engaged in many a hero's journey of consciousness that created great human civilizations on the one hand and gave us a glimpse of human potential on the other.

When the agricultural era was replaced by the industrial era, the hierarchies imposed by the rational mind gradually gave way to a more equitable sharing of meaning processing through the advent of such great institutions of human thought as democracy and capitalism. Thus began the "age of enlightenment," and a rational worldview began its journey to become the dominant worldview. Finally, with the advent of technological societies, when women forced their way into meaning processing, thanks to the women's liberation movement, the rational worldview completed its domination.

Returning to the Brahmins of the rational stage: A Brahmin at this stage may begin to overcome the inertia of the rational mind by pursuing an intense curiosity in some narrow field of knowledge. And an occasional foray of fundamental creativity in a narrow field begins a creative life. Again, the tendency is to develop a narrow field of expertise. The Brahmins of the rational mind take quantum leaps to the supramental to tackle primarily problems of the "outer" arena; in other words they participate in outer creativity and are still fascinated with the outer world and manifestations in it. Even if such people get an insight about the nature of love, they would write a novel or a song about it or write a scientific paper on love, depending on their narrow window of expertise. But they would not be inspired to consider living the insight for their own personal transformation.

A few Brahmins, however, get bored with outer expressions of creativity and see the potential of the creative journey for inner exploration. They set their creativity toward the study of higher states of consciousness and achieve the states of pure awareness called *savikalpa samadhi* (samadhi with subject-object split) or *satori*. Even so, they would not particularly attempt to transform their ego-base, nor would they attempt to integrate emotions in their way of life. So even this is outer creativity of a sort; the product is an accomplishment of the ego. Naturally, people of this stage, when they attain samadhi, declare themselves enlightened, illumined. Hence Sri Aurobindo has called this level of mind an "illumined" mind.

A number of great illumined minds showed up in the agricultural era, and the number has clearly increased in recent times. A few people of the illumined mind have even journeyed on the next stage of the path, transformation, at least partway. However, as a society, we are nowhere close to developing the next stage of human evolution—the intuitive mind, in which the mind is preoccupied with the supramental. Why is that?

I think the dominant reason is that the emotions have never been properly integrated in our world culture, which remains lost in the illusion of rationality while being quite incapable of dealing with the "animality" of instinctual emotions. This difficulty is further discussed in chapter 21. Until the emotional vital mind is fully integrated with the rational mind, we are evolutionarily stuck.

The Intuitive Mind. After the rational mind integrates the emotions with thinking, a genuinely new class of mind will evolve—the intuitive mind. This is a specific prediction of the new biology.

At the level of intuitive mind, people will intuit that without some transformation from their basic ego level of being, it will be impossible to shift the emphasis of life from the mental to the supramental. Naturally, they will use the insights obtained in their forays to the supramental to transform themselves to higher levels of being. In other words, people of this level fully participate in inner creativity; they manifest their supramental insights about the contexts of mental and vital functioning in their lives. Because emotions are now fully integrated with reason and meaning processing, and because feelings and meanings are treated on an equal footing (always with the proper supramental context as background), people of the intuitive mind will live life mostly with positive emotions.

These people therefore will not remain intellectually frozen in their adult lives. Instead, they will continue to explore the supramental to discover the archetypal themes of life to a fuller extent. They will first make mental representations of these themes, but then they will live them. They will "walk the talk" and develop the new brain circuits that will make walking this way relatively effortless.

There are already a few such beings in our midst, as has been the case for millennia. They are the people who create and sustain our civilizations: Socrates, Buddha, Jesus, Muhammad, and the like. Recent examples of such great minds are Henry Thoreau, Mahatma Gandhi, Eleanor Roosevelt, Desmond Tutu, and the current Dalai Lama. These are people who regularly have peak experiences of creativity and spirituality; who are able to love unconditionally more often than not, especially when occasions demand it; who maintain more or less a state of equanimity or environmental independence. They are by no means perfected beings, yet they maintain a sense of humor about their failings. These partially transformed people are of the type psychologist Abraham Maslow (1971) studied and called "people of positive mental health."

At the highest level of mental evolution, as the intuitive mind matures, it evolves toward what Aurobindo called the "overmind" and what Jung called the "individuated" mind. People of the overmind have minds of the highest possible mental perfection.

The concept of reincarnation is relevant here. Biologists do not much bother with the issue of survival after death or reincarnation but they should, for three reasons. First, until and unless they do, a reconciliation of biology with religious and spiritual traditions will be hindered. Second, the case for reincarnation, in both theory and data, is now very strong (Goswami 2001; see also chapter 18). Third, the theory and data in favor of reincarnation provide some of the best evidence for accepting a science within consciousness.

Reincarnation is important in the context of intuitive mind because ours is a learning journey that requires many lives. The purpose of the journey is to learn the supramental archetypes, or how and why the world works. When all archetypal themes of life are discovered and lived, then a person has no need to reincarnate again to have another mental life. Therefore, such a person can be said to be liberated from the birth-death-rebirth cycle.

Liberation can be reached in two ways. The first is by making brain circuits through discovering and living all the archetypal themes relevant to life. Sometimes this is called the discontinuous method of reaching liberation, because the discovery process for the archetypal themes always involves quantum leaping. But there is also continuity in approaching the problem. The themes are tackled step by step, one by one. Achieving overmind in this fashion is also how Jung defined his concept of individuation. People of overmind are individuated—they have exhausted all the archetypal themes of life during their reincarnational journey.

In the second, continuous path to liberation, one arrives, through surrender of the ego, in states of creative sleep (states of creative unconscious processing, called *nirvikalpa samadhi* in Sanskrit) during which one is in God-consciousness. When one has effortless access to such creative processing whenever in the unconscious, then quantum leaps to truth and the making of brain circuits by living the truth are no longer necessary. Appropriate action requiring quantum leaps follows automatically, easily, without effort, flowing from the God-consciousness these people live whenever unconscious.

Beyond the overmind is the supermind. To have supermind is to develop the capacity of directly representing the supramental onto the physical. This we cannot do in the mental age of evolution, with the brain. Supermind is our next and final (in the current reckoning) stage of creative evolution.

I have one final question to pose before ending this chapter. Will the further evolution of the mind be accompanied by a concurrent evolution of the brain, the representation-making apparatus? This very interesting question warrants research. Science fiction authors sometimes depict advanced extraterrestrial humanoids with bigger brains and shrunken bodies (recall the movie character ET). With evolution of the mind, the brain's representation-making capacity must improve, but it is not a question of more content requiring a greater brain size. (Incidentally, a physically bigger brain would require a simultaneous evolution of the size of the pelvic passageway in the human female.) The idea of shrunken bodies comes from the materialist notion that we will more and more depend on machines to do our physical chores; simple nonuse will shrink our bodies, and the change will be fixed through evolution. Biology within consciousness predicts a different scenario. As you have seen, in the new biology, the future of evolution consists of an integration of the processing of feeling and meaning, body and mind.

PART 6

MORE NEW BIOLOGY: The SUPRAMENTAL DIMENSION

18

The BIOLOGY of DEATH and SURVIVAL after DEATH

What is death? I first entertained this question when I was very young. I was looking out the window of our house at some hoopla happening on the street and getting more and more curious. Four people were carrying something jointly on their shoulders, and quite a few others had joined them in what seemed to be a chant. The words of the chant translate to something like "the name of Rama is the real thing."

When next I saw my mother, I told her what I had seen, and she explained what it was: A dead body was being taken to the burning ghat. That's when I asked my question: "What is death, Mother?"

"Death is a passing away to the next world," said my mother, adding, "We all die."

The answer created more mystery, so I asked, "What passes away, mother?"

"The soul, of course," said Mother, adding, "the real you." After that she went away in a hurry, leaving me with yet another mystery. It took me many years to find a satisfactory answer.

My mother's reply still resonates with people whose world-view is connected to a spiritual tradition. The majority of mothers today, though, having grown up in a more materialist culture, would answer the question "What is death?" quite differently. A mother might say, "It depends. There is brain dead, heart dead, and cell dead." And then she might explain each kind of death. "Brain death occurs when the brain no longer functions. This shows up as a flat line on an EEG monitor. Heart death is when the pulse stops. However, the heart is just a blood pump and the pumping can be maintained through artificial means, indefinitely. Cell death takes place when the individual cells stop functioning properly and the organs begin to decompose."

Today's mechanically sophisticated children may actually understand and appreciate such an answer. A few will inquire, "What happens to *me*, Mom, when I die?" Then the modern mother may be in a fix. The nihilist answer from materialist science—"There is no me and nothing survives death"—doesn't satisfy, and the modern mother knows that. So either she will retreat or switch to a more spiritual point of view and say something about a surviving soul.

The question "Is there a surviving soul after death?" is very much alive even after decades of rampant materialism in our culture. For this reason, doctors and hospital administrations are ambivalent about whether or when to terminate the life support systems of brain-dead patients in prolonged coma. How can science determine when the soul leaves the body?

Many scientists believe that the questions of death and especially of survival after death are metaphysical and are not answerable or tractable in science. This view is rapidly changing, however. This change holds a real opportunity for bridging the scientific and spiritual positions about the question of death and survival that should interest all biologists.

When we base our science on the primacy of consciousness, as quantum physics demands that we do, then the question of survival takes a new turn. With death the material body does not function any more and the entropy law takes over—the result is decomposition. However, consciousness and its other possibilities—vital, mental, and supramental—remain intact; no entropy law exists in these domains.

The intriguing question then becomes "Is there any 'me' in what survives the death of the physical body?" Let's examine this question in some detail.

For the staunch materialist neuroscientist, what forces this issue to the forefront of neuroscience is the enormous body of empirical data on what are called near-death experiences that have accumulated in the past few decades.

NEAR-DEATH EXPERIENCES

Some people who have a cardiac arrest literally die (brain death as shown by a flat EEG reading), only to be revived a little later through the marvels of modern medicine (Sabom 1981). Some of these near-death survivors report having witnessed their own surgery as if they were hovering over the operating table, an experience called *autoscopic vision*. They are able to give uncannily specific details of their operation, leaving no doubt that they are telling the truth, however difficult it is to rationalize such experiences through materialist neuroscience. They are not seeing anything with their local eyes, with local signals, that much is clear. Indeed, even blind people report such autoscopic vision during near-death coma (Ring and Cooper 1995), so there can be no doubt that local signals and local eyes are not involved. These near-death survivors are seeing using their nonlocal distant-viewing ability, tapping into the eyes of the doctors, nurses, and others involved with the surgery (Goswami 1993).

Even this explanation presents problems. How can they see even nonlocally while they are brain dead, unconscious, and quite incapable of collapsing possibility waves? The answer is unconscious processing, of course. A chain of uncollapsed possibilities can collapse going backward in time, as verified in the delayed choice experiment (see chapter 6). For the near-death survivor, the "delayed" collapse takes place at the moment brain function returns as noted by the EEG; the return of brain function precipitates the whole chain of retroactive collapses going backward in time.

Many near-death experiences include another aspect that illustrates the main issue in this discussion. Reports by near-death survivors include not only autoscopic visions but also out-of-body experiences, specifically, going through a tunnel to meet dead relatives and even beings of light. Where does the vivid out-of-body feeling come from? These patients, being brain dead, no longer have any identity with their physical body, but they do identify with their mental and vital bodies, their "soul." These out-of-body experiences are direct proof (that passes the criterion of weak objectivity) of the continuation of "I" after death. The "I" remains in the mental (and also vital) bodies.

Other direct data for survival of an "I" also exist but are more controversial, for example, data on psychic channeling, a phenomenon in which a medium "channels" or transmits messages from a disembodied "soul."

A THEORY OF
SURVIVING INDIVIDUALITY

In chapter 2, I referred to the problem of individual ego identification and stated that it arises from past conditioning. Let me elaborate some details of how that conclusion is reached.

The physicist Mark Mitchell and I (Mitchell and Goswami 1992) studied the quantum mechanics of the repeated measurement of the position of an electron and found that since any measurement involves memory, subsequent measurements involve not only the processing of the present measurement but also of the past memory. The feedback of the past memory gives rise to a nonlinearity in the otherwise linear quantum equation of movement (the Schrödinger equation) of the electron. This nonlinearity leads to a modification of the probability of the electron's response to the new measurement in favor of its past measurements. In this way the electron becomes "conditioned" to show up in the same position in which it showed up before. That is, it behaves more and more like a classical Newtonian object, forgetting its quantum freedom.

When this result is generalized to the brain and its making of representations of mental meaning, we must conclude that reflection in the mirror of our past memory has the same effect of producing conditioning: We tend to respond to a previously experienced ("learned") stimulus as we have responded before. This tendency produces a pattern of habits of mental response. Even the archetypes that originally set the contexts of our mental thinking become personalized, an aspect of ourselves that we call our character. This mental habit pattern plus our character plus our "his"tory (or "her"story) is our ego.

The mind, as Descartes knew, has no micro-macro division, no structure. So purely in terms of structure, there is no such thing as an individual mind. However, the conditioned character response pattern that we each identify with does give us, functionally speaking, an individual mind.

Now the crucial point can be made about survival after death. When we die, the brain dies, with its memory of personal history. But the memory that is represented by the character habit patterns remains safely locked into the appropriate nonlinear quantum equation of movement that comes into play whenever these habit patterns are used. This kind of memory I call "quantum memory" (Goswami 2001).

Obviously, then, quantum memory survives the death of the physical body. The surviving quantum memory of the mental (and similarly the vital) bodies can be recognized in what the spiritual traditions call the soul. Or at least this is one connotation of the word *soul*.

The phenomenon of channeling mentioned earlier provides us with direct proof that the disembodied soul consists mainly of mental habit patterns and character in the sense just described. In a project that yielded very compelling data, the parapsychologists Gilda Moura and Norman Don (1996) studied a channeler who channeled an entity named Dr. Fritz, who had been a surgeon. The channeler did not have any surgical skills, yet when he channeled the entity he was able to perform complicated surgical operations, often using primitive tools. To make the study conclusive, Don and Moura measured the channeler's brain waves while in the ordinary waking state and also while he was channeling. During channeling, the channeler's brain waves consisted of very high-frequency beta waves at over 40 Hz. In his ordinary waking state, the channeler's brain waves always conformed to the more usual beta-wave frequency, much below 40 Hz. This kind of channeling experiment shows clearly that the channeler's own vital and mental habit patterns and character exhibit changes during channeling that can only be attributed to the habit patterns and character "borrowed" from the disembodied channeled entity.

REINCARNATION

Materialist biologists sometimes worry that "certainty about the next life is simply incompatible with tolerance in this one," as the writer Sam Harris said. I think there is some truth to this concern. Indeed, too much concern about the other world has made certain religions especially negligent about this world. Such concern is misplaced, because the data speak in favor of not only survival after death but also rebirth. You cannot escape this world!

In fact, ever more striking and very solid data exist for the survival of a "soul" consisting of the functional vital and mental bodies of a person in the phenomenon of reincarnation—rebirth of the surviving soul in a new physical body. Compelling data for reincarnation were collected by the psychiatrist Ian Stevenson at the University of Virginia Medical Center (Stevenson 1974, 1977, 1987). The data consist mostly of past-life recall accounts by children of various nationalities, both East and West, that were carefully followed through and verified. In one account, the child described some money hidden in a secret compartment in a house that he recognized as having been his house in a past life. The money was found just as the child described. It was also verified that the child had never been to that village, let alone to that house.

This kind of data indicates, at the least, a nonlocal correlation between a present human being with one that lived in the past. Suppose that by using this nonlocal correlation, a future incarnate body uses the character and habit patterns developed by the incarnate body that lived in the past. Then that future body can rightly be called the reincarnation of the past person, can't it? This transmigration of (mental and vital) character and habit patterns then gives us a model of reincarnation that can be verified by examining whether character traits are inherited by people that cannot be explained otherwise.

Indeed, Stevenson's data show many examples of the inheritance of such acquired characteristics through reincarnation, examples such as geniuses, people with unexplained phobias, people with unusual language ability, and so on. Details of such data are given elsewhere (Goswami 2001).

One final comment on this topic: Reincarnation is often thought of as a concept of Eastern religions, but this view is mistaken. Reincarnation was very much a part of Christianity until the fifth century AD, when the idea was eliminated from official Christian teachings for political reasons. Even so, as the philosopher Geddes MacGregor (1978) has cogently argued, Christianity has indirectly retained the idea of reincarnation in such notions as purgatory, a place where one waits after death and learns.

THE INTEGRATION OF
SCIENCE AND SPIRITUALITY

One of Nietzsche's characters famously declared, "God is dead." Nietzsche was a little premature, I think. But it is fair to say that an assassin made a serious attempt to murder the idea of God, without quite succeeding. Of course the assassin was not a person, but another idea. This idea can easily be identified as Darwinism. Ever since Darwin, science and spirituality have been antagonistic to one another. Can this antagonism give way to a much-needed integration?

The 1960 film *Inherit the Wind* revolved around the conflict between creationism and Darwinism over the teaching of Darwinism as part of biology in schools. In the final scene of the movie, a lawyer holds up a copy of a Darwinian textbook in one hand and a Bible in his other hand. He then seems to weigh them in the air and balance them against each other. This lawyer has been defending a schoolteacher for the crime of teaching Darwinism in his classroom. Today, it would be the other way around; a lawyer would have to defend a teacher for teaching intelligent design in a classroom. But what to me was interesting in the movie was the last message it gave: The lawyer looks at the two books, smiles, puts them side by side in his bookcase, and exits the courtroom. The message was that yes, science and spirituality can coexist.

With biology within consciousness, we can do better. We can integrate science and spirituality; we can integrate the dichotomy we have lived with ever since Darwin. With scientists investigating and incorporating within science ideas such as survival after death and reincarnation, even the last vestige of resistance against the integration of the two great traditions should disappear.

19

A New Bioethics *and* Deep Ecology

The notions of ethics were discovered by the great mystics early in the era of the rational mind, mystics whose teachings about ethics became integral to all religions. I'm speaking here of religious ethics—be good; do good; do unto others what you would have them do unto you; love everyone.

Religions maintain that, like physical laws, ethics is compulsory. If you do not follow ethics, you suffer. If not now, later, maybe after death. In some religions, this conclusion gives rise to the concept of hell.

Few philosophers take the idea of hell seriously, but they do see the usefulness of ethics in our society. When scientific materialism came into the picture and religion no longer seemed viable, the philosopher Immanuel Kant tried to establish that ethics is a categorical imperative. However, it is generally agreed that Kant failed to show that ethics is compulsory in the sense that physical laws are compulsory. You seem to be able to get away with ethical violations (e.g. the politicians of the world), but you can never defy the law of gravity. I have shown elsewhere (Goswami 2008) that invoking reincarnation solves this problem of whether ethics is compulsory. You can get away from the consequences of violating ethics now, but Captain Karma eventually gets you. You have to learn to be ethical, or else you get forever caught up in the birth-death-rebirth cycle.

Biologists, that is, Darwinists, have invented a form of scientific ethics based on biology, specifically, on the principle of genetic determinism. The idea is that our behavior is entirely determined by our genes; we are gene machines. Our mind, consciousness, and macrolevel behavior all have one ultimate purpose, to perpetuate our genes, to guarantee their survival. It follows from this perspective that we should have some interest beyond selfishness, some natural tendencies of selfless, altruistic behavior. For example, if I have some genes in common with another person, my genes would naturally gain propagation and survival advantages by causing me to take care of this other. Hence, I behave altruistically toward another based on how much genetic commonality I have with that person.

This theory is a good one because it's testable. If the theory were correct, this kind of bioethics would be compulsory—our genes would make sure of that. Unfortunately, the empirical data on altruism just do not conform to this theory. Empirically, altruism extends way beyond our genetic connections.

We have still another way to look at ethics—evolution. One problem remains with all the ethical philosophies discussed above: Why do some people follow ethics and others don't? In the old days, fear of hell or desire for heaven was an incentive. These days, who takes heaven and hell seriously enough to sacrifice selfishness because of them? Still, so many of us, even today, try to be "good" in our daily living, even in the face of the explosion of unethical behavior in our societies. Why? I submit that the reason is evolution. An evolutionary pressure is present that some of us experience as a calling, and we respond.

It follows that ethics need not be looked upon as religious or spiritual, nor is there any need to compromise by adopting a materialist ethics of the greatest good for the greatest number, or a bioethics driven by the genes. We can solidly base ethics on the very scientific notion of evolution.

Let's define an evolutionary ethics: *Ethical actions must maximize the evolutionary potential of every human being.*

Once we become established in an evolutionarily ethical relationship to all our fellow human beings, it is time to ponder our ethical responsibility to all creatures, great and small, including the responsibility to our nonliving environment. In short, let's ask, What is our responsibility to the planet Earth, to Gaia?

The phrase deep ecology was coined by the philosopher Erne Ness, and ever since it has influenced New Age thought (read Devall and Sessions 1985). The idea is to replace the shallow, Cartesian ecology of "human over nature" with a deep ecology of humans being a part of living nature and being interconnected with it. If we acknowledge our interconnection with nature and learn to respect and live it, then we will not pollute and destroy nature in our vain search for happiness. In other words, we are giving an intrinsic value to preserving nature.

But if we abandon the idea of free will in favor of materialism and the laws of physics, then whence comes the intrinsic value for preserving nature? Deep ecologists depend upon the philosophy of holism. The interacting whole is greater than the parts; in such a whole an intrinsic value, even a nature spirituality, may emerge, or so the deep ecologists hope.

It is easy to see, however, that a deep ecology based on holism is not deep enough. In holism-based deep ecology, there may very well be an emergent quality of interconnectedness that we can identify as nature-spirit (Capra 1982), but can we live in harmony with it? Can we transform from our selfish ways, from our genetic and environmental conditioning? Even in emergent holism, we don't really have the power of downward causation, only an appearance of free will, because the causal forces of upward causation are hidden from us. If our intentions are without causal efficacy, how can they bring about a transformation to living integrally with our ecological community?

For example, in one model of emergent holism, the interconnectedness is seen as an order within chaos (Prigogine and Stengers 1984). The interconnectedness of emergent holism is shallow because it is an interconnectedness proposed on the basis of local interactions (which die off with distance) among the interacting parts of the whole system.

In contrast, the physicist David Bohm (1980) went one step further to propose a nonlocal interconnectedness based on quantum nonlocality. In Bohm's "causal" interpretation of quantum physics, the wave-particle relationship is seen a little differently. In this approximate reformulation of quantum physics that makes it deterministic (hence causal), the wave is seen to guide the particle through a nonlocal "quantum potential." The quantum potential does not die off with distance; hence the particle can exhibit bizarre non-Newtonian, nonlocal behavior. A network of interacting quantum systems, in Bohm's model, would have a potent interconnectedness via the hidden order of the underlying waves. In Bohm's terminology, such interconnectedness is called the *implicate order*, while the particle order of manifest beings is called the *explicate order*. The explicate order may look disconnected except on occasion (as in the experiments with quantum correlation that gave away the secret of quantum nonlocality). Underneath there persists that deep interconnectedness through the implicate order. Being nonlocal, this interconnectedness is deeper than that proposed by theories of emergent holism.

A static example of Bohmian interconnectedness is a hologram. You have seen a hologram no doubt, perhaps on your credit card. A hologram is made by capturing wave interference patterns from an illuminated object, and each little captured part of the wave interference pattern stores all the information of the whole object, albeit with poor resolution. Analogously, the objects of the explicate order may appear as isolated individuals, but they retain the nonlocal potency of the interconnected whole of the implicate order. The neurophysiologist Karl Pribram, the systems theorist Ervin Laszlo (2004), and the former astronaut Edgar Mitchell all have endorsed this kind of deeper ecology. But once again, is this deep enough?

The answer is still no, because these theories are deterministic. They have no mechanism for transformation by which we— Newtonian/sociobiological/behaviorally determined and ecologically destructive creatures that we are—can change our ways. A deep ecology that fulfills Ness' vision will describe not only interconnectedness but also causal potency for transformation.

This causal potency for transformation is ours to claim as soon as we claim our capacity for operating in the nonordinary state of quantum consciousness. Our destructive tendency comes from our genetic, instinctual, and psychosocial conditioning. When we rise above our conditioning even temporarily, we can change: We can be harmonious with our ecosystem, with our whole planet, with Gaia. We can do it not by wishing, nor by philosophy, but only through intention and the creative process, that is, by taking a discontinuous quantum leap. This leap is the missing element in the deep ecology movement. Deep ecology requires not only abiding by a few rules for preserving our ecosystem or passing a few governmental laws preventing environmental pollution, but also taking actions in ambiguous situations that demand a creative quantum leap.

When you take such a quantum leap you realize one astounding thing: *I choose, therefore I am, and my world is.* The world is not separate from you. When we do this en masse, we will leap into a truly Gaia consciousness.

20

LAMARCKISM *and the* MYSTERY *of* INSTINCTS

I have shown throughout this book that Darwinian mechanisms of evolution are too limiting and that biology must embrace a far broader understanding of inheritance. To make this shift, we must revive Lamarckism by reinterpreting it in light of quantum creativity. We'll soon see how this neo-Lamarckism can explain yet another mystery of biology: the development of instincts. Further, as will become clear in the next chapter, this new way of understanding instincts could, in its turn, be essential in getting humanity out of our current evolutionary doldrums.

According to Lamarck, new species evolve from ancestors through the inheritance of acquired characteristics. However, "history dealt unkindly with Lamarck," as Francis Hitching remarked (1982). The reason lies in a misunderstanding of what Lamarck meant by acquired characteristics. The usual objection is that, for example, a bodybuilder's child doesn't necessarily have big muscles, and if a cat loses a hind leg in an accident, her kittens are not born sans that same hind leg.

However, these are not good examples of what interested Lamarck. Rather, he believed that the use and disuse of an organ, as influenced by the environment, can affect evolution. "The mole, whose habits require a very small use of sight, has only minute and hardly visible eyes," he wrote; this change, Lamarck believed, is the result of permanent disuse of the eyes. Consider also that we get thick skin on the soles of our feet if we always walk barefoot; the ostrich gets calluses on its knees where it kneels. This type of change in response to environment is the point of Lamarckism.

Recall that the chief objection to Lamarckism is that information moves in the "wrong" direction, that is, from proteins to genes, violating the "central dogma." Let's look at a hybrid Lamarckian-Darwinian argument that tries to dodge this problem, describing how an environmentally influenced trait might spread in a population without (initially) involving the genes. Hitching (1982) recounts a three-stage process put forward by the biologist John Maynard Smith and summarized here:

- Stage 1: Adaptation to environmental stimuli during development. Callused knees in ostriches, for example, arise during the juvenile stage because of new environmental stimuli, such as drought, that make the ground on which the ostrich kneels very rough. (For other traits, the adaptation might begin earlier, even during embryonic development.)

- Stage 2: Canalization of development. If the draught continues for several generations, the developmental response of callused knees becomes deep and established in the bulk of the population. This effect is called *canalization*.

- Stage 3: Genetic assimilation. After some generations, the evolutionary pressure of the environment itself will bring about the selection of gene mutations that (maybe somewhat gradually) lead to the same canalized development. At this stage, the response occurs even without the environmental stress; we can say that genetic assimilation has taken place.

I maintain that Maynard's scenario is not credible. It assumes canalization of the acquired characteristic over virtually the entire species. This effect would required prolonged environmental stress lasting for many generations, a highly unlikely scenario. Furthermore, the idea of genetic assimilation through Darwinian step-by-step gradualism does not seem credible. The allotted time is too short. Is there an alternative? Yes.

We already know a credible mechanism for genetic assimilation: creative evolution. When the entire gestalt of gene variations is available in possibility in a way that synchronizes with the canalized form-making, consciousness chooses the entire gestalt simultaneously.

On the issue of canalization of an acquired characteristic over an entire population, can we find a cause more credible than long, drawn-out environmental stress? The answer is yes, provided we overcome an implicit prejudice of biology: that the genes are the only hereditary system!

What If Genes Aren't the Only Mechanism of Heredity?

Some biological materialists are reluctantly acknowledging new experimental results that show the existence of nongenetic inheritance systems (Jablonka and Lamb 2005).

One of the first such experiments was carried out by the immunologists Edward Steele and Reg Gorczynski (Steele 1980). In these experiments, the immune system tolerance of mice was altered by a special technique. Surprisingly, these authors found that the tolerance is acquired by subsequent generations of mice.

If young mice, soon after they are born, are injected with a large quantity of foreign cells from other mice (about ten million cells), their immune system, not having fully developed, does not reject the foreign tissue. Apparently, the genes in somatic cells of the injected mice are changed so as not to respond to foreign cells, a condition they retain through all their regenerations within the injected mice. Amazingly, this immunological tolerance is passed on to subsequent generations, demonstrating clearly the validity of the principle of Lamarckian inheritance of acquired characteristics.

There was no doubt that a change in the genes was involved. "When you examine the mouse's children and grandchildren, you find the tolerance to foreign cells passed on in just the numbers the conventional Mendelian theory would predict—that is, half the children, one quarter the grandchildren," said Steele.

What is the mystery that allows information contained in somatic cells to be passed on to germ-line cells? Steele suggested that in a chaotic situation like the early stages of formation of the immune system, viruses may pick up copies of modified RNA and carry them to the germ line. According to the central dogma, the amino acid sequence in a protein can never be transferred to a DNA, but in the virus explanation, the modification occurs because the DNA is replicated off the modified RNA "template." This type of experiment eventually gave rise to the idea of "epigenetic inheritance systems" (Jablonka and Lamb 2005). The notion is still controversial because no evidence exists for epigenetic hereditary information that is evolutionarily adaptive. Nevertheless, this work suggests one mechanism for bypassing the central dogma.

However, because of their materialist prejudice, biologists have been missing another, much more versatile source of hereditary information transfer besides genes—the nonlocal morphogenetic fields.

HEREDITARY INFORMATION TRANSFER
THROUGH MORPHOGENETIC FIELDS

Changes in the environment affect the cells and even conglomerates of cells (i.e., the organs). Further, by affecting cells and organs, environmental changes also affect the nonlocal morphogenetic fields correlated with these cells and organs, because the fields must adjust to changes in their representations in the physical body.

In making such adjustments, consciousness has three options for altering the vital body. If the adjustment to the environment requires a moderate change, a solution may be found within the extant repertoire of morphogenetic fields in the vital body of the organism. In other words, situational creativity (vayu) is sufficient. If on the other hand a major change is called for, fundamental creativity (tejas) must come into play. If the adjustment required is so small that the organism can get by without making changes, then conditioning (inertia, or ojas) prevails.

In this way, persistent environmental changes to the soma of an organism produce changes in its morphogenetic fields that become fixated as part of the repertoire of these fields. At first, the repertoire is individual, but the nonlocality of consciousness in the vital arena may assure canalization for the entire species, perhaps subject to a threshold effect (see below). Subsequently, when appropriate quantum possibilities of gene variations are available coincidentally with the appropriate environmental pressure, consciousness, through biological creativity, can collapse them into operation to achieve synchrony of form with the previously modified morphogenetic fields. At this point, assimilation has occurred.

HEREDITARY INFORMATION TRANSFER THROUGH REINCARNATION: GROUP SOUL

The sensitive reader will see the analogy with reincarnation here (see chapter 18). However, there's a twist. In ordinary human reincarnation, for which we have ample favorable data, the mental and vital propensities of an individual (technically called the *quantum monad* and popularly called the soul) survive death and are reused by a incarnate body in the future through a special non-local correlation. However, only beings living at the level of rational mind (see chapter 17) undergo this type of reincarnation.

For beings living at the level of vital mind, the mental individual ego structure has not yet developed. Thus, nonlocal consciousness dominates the weak individual consciousness. This is true particularly of animals but may even have been true of early humans before the evolutionary stage of the rational mind. In spiritual traditions that teach a belief in reincarnation, this idea is expressed by the notion that animals have group souls, a concept similar to species consciousness (see chapter 12).

In practical terms, this nonlocality means that one individual's quantum monadic propensities can reincarnate in several individuals at once. That is, many later individuals may have the nonlocal correlation that permits them to tap the characteristics of a particular previous quantum monad. Thus, individually learned vital propensities, as reflected in the modification of the morphogenetic fields, can propagate rapidly throughout the species. Finally, a threshold is reached when the canalization of the entire species is achieved.

The informed reader may have noticed that the neo-Lamarckian theory that I propose here is quite in tune with the ideas of the biologist Rupert Sheldrake (1981). This is not a coincidence. The present theory incorporates the earlier ideas of Sheldrake.

This kind of fully Lamarckian evolution can explain such characteristics as the horned knees of the ostrich and the thickened soles of human feet. In addition, and most importantly, this neo-Lamarckian vital mechanism bypasses the central dogma and opens the door for the explanation of a very enigmatic phenomenon: instinct.

Neo-Lamarckism and the Explanation of Instincts

Abraham Maslow said, "It is tempting, if the only tool you have is a hammer, to treat everything as if it were a nail." For biologists, nowhere is this tendency more apparent than in their habit of thinking of behavior in terms of genetic determinism. "All bodily structures and functions, without exceptions, are products of [genetic] heredity realized in some sequence of environments. So are all forms of behavior, without exceptions," said the biologist Theodore Dobzhansky. Some behavior, yes; genes affecting behavior through production of hormones, yes. But *all* behavior? A resounding *no*. Genes are basically instructions to build proteins, and the relation of the genotype and phenotype is so complex that except for a few cases of one-to-one correspondence, it is hard to claim that there is a genetic explanation of the functions and behavior of a trait.

Among the behaviors for which a solely genetic explanation is impossible to concoct is instinctual behavior. On the other hand, the very definition of instinct is a behavior that is present without having been learned. How else could it be present, then, if not by inheritance through genes? This is the mystery of instincts.

To make the point clear, consider a few examples. A chick still inside the eggshell tries to get out by pecking. Not only that, it knows not to peck at random but just at the part where there is an air space. You cannot call that learned behavior, because the behavior shows up even before chicks see daylight. Or consider the following tale of instinct told by the geneticist Henry Fabre. A certain species of beetle lays eggs within the body of a tree. When the eggs hatch, the sluggish larvae called grubs spend three whole years within the tree, tunneling through the wood. When they emerge from the tree at last, they build a nest. The complicated nature of the nest leaves no doubt that the instructions could not have come from any genes.

What is the explanation for instincts? Darwinians do not give up easily. Consider the following tale told by the biologist Douglas Spalding, a contemporary of Darwin, and retold by the biologists Eva Jablonsky and Marion Lamb (2005), about the Darwinian explanation of the evolution of instincts:

> Suppose a Robinson Crusoe to take, soon after his landing, a couple of parrots, and to teach them to say, "How do you do, sir?"—that the young of these birds are also taught by Mr. Crusoe and their parents to say, "How do you do, sir?"—and that Mr. Crusoe having little else to do, sets to work to prove the doctrine of inherited association by direct experiment. He continues his teaching, and every year breeds from the birds of the last and previous years that say "How do you do, sir" most frequently and with the best accent. After a sufficient number of generations his young parrots, continually hearing their parents and a hundred other birds saying "How do you do, sir?" begin to repeat these words so soon that an experiment is needed to decide whether it is by instinct or by imitation; and perhaps it is part of both. Eventually, however, the instinct is established. And, though now Mr. Crusoe dies, and leaves no record of his work, the instinct will not die, not for a long time at least; and if the parrots themselves have acquired a taste for good English the best speakers will be sexually selected, and the instinct will certainly endure to astonish and perplex mankind, though in truth we may as well wonder about the crowing of the cock or the song of the skylark. (Quoted in Jablonsky and Lamb 2005, 286–87.)

Jablonsky and Lamb (2005) elaborate further:

> In Spalding's thought experiment, the selection of Mr. Crusoe's parrots was initially artificial. However, it is not difficult to see how, even without human involvement, behavior that was at

first learned could become innate through natural or sexual selection. (288)

According to Jablonsky and Lamb, animals may learn a new, risky behavior such as avoiding a new predator when there is an urgent need to learn, in other words, when the learning has survival and selective advantage. Gradually, the learning ability improves and the behavior becomes canalized. After many generations, the learned response becomes instinctive, for all practical purposes. Eventually, genetic assimilation occurs.

This theory, as you can see, is not different from the one previously quoted from Maynard Smith about the Darwinian explanation of the inheritance of acquired characteristics. However, my previous criticisms stand, namely: (1) the availability of prolonged environmental stress to produce canalization over an entire species population is unlikely; and (2) the feasibility of producing genetic assimilation through a Darwinian step-by-step process in the relatively short time available is questionable. As before, I posit that the learning becomes canalized because the modified morphogenetic fields are nonlocal and thus can have effects in building form beyond the originally altered individual. The reincarnational inheritance of the modified morphogenetic fields (and the resulting modified forms) by an entire group facilitates learning of a behavior in future generations and spreads the learning in the bulk of the population. Eventually that learning becomes so quick as to be virtually instinctual. Finally, genetic assimilation via biological creativity then makes the behavior fully instinctual.

CHAPTER

21

$\mathcal{E}$VOLUTION:
The $\mathcal{N}$EXT $\mathcal{S}$TEP

*T*he future of evolution, the omega point, is that humans will develop the capacity for representing the supramental in their physical bodies. Naturally we want to ask how and when that will come about. It is a little premature to engage with that question, unfortunately. A more pertinent question is, What is the immediate future of our evolution? I said in chapter 17 that we are in the mental age of evolution, with one more stage to accomplish: the stage of the intuitive mind. I also said that, evolutionarily speaking, we are stuck. How do we get ourselves unstuck? We must integrate our feeling (vital) mind with our rational mind. How do we evolve in that direction? Can we? These questions are the subject of this final chapter.

Someone once said, "Man has learned to fly like a bird in the sky; man has also learned to swim like a fish in the ocean; but, alas, man has yet to learn to walk like a human being on this earth." What does it mean to be fully rational, let alone fully human? Human beings began with a tremendous leap in the capacity for making representations of the mind. A major privilege of being human is in developing the rational capacity fully.

As the philosopher Ken Wilber emphasizes, however, at each level of the great chain of being, the previous level must be integrated. Only when humans have integrated the vital mind can we be said to fulfill our full potential for rationality.

Have we fully integrated our vital body in our mental being? In view of the current state of mind-body disease and healing (see chapter 15), we are far from it. We express emotions and suppress emotions, but we haven't learned to meditate on them. We share the base instinctual emotions with animals. We are so insecure, so prone to succumb to these base instincts, that we sometimes suppress even the noble emotions associated with our higher chakras, beginning with the heart.

EMOTIONAL INTELLIGENCE AND INSTINCTS

Lately, people have been talking a lot about "emotional intelligence" and "emotional maturity" (Goleman 1995), but most of the discussion begins from a behaviorist and materialist point of view and is thus very incomplete. Without including the vital body, we cannot even define emotion or feelings properly, let alone realize that emotions are effects of feelings on the mind and the physical body.

What are the signs of emotional maturity? One major sign is the ability for frequent engagement with positive emotions such as love and humor. Another sign is the ability to remain unperturbed by a negative emotional environment (an angry crowd, for example). As the psychologist Uma Krishnamurthy (2008) says, emotions are contagious, more contagious than bacteria and viruses. This is particularly true of base or negative emotions. Thus, an emotionally mature person has developed a sort of immunity to base emotional contagion.

Freud recognized long ago that two of our base emotions, sexuality and (emotional) ego are instincts. Though they later came to be called drives, I think sexuality and emotional ego really are instincts. They are wired in the limbic brain, so that whenever they are aroused, helpless action follows. The spreading of negative emotion is facilitated by helpless, automatic actions; in this mode, individuality easily gives way to a group consciousness. I think other negative emotions, such as fear, insecurity, greed, competitiveness, and jealousy, have also become instincts because the group consciousness they induce facilitates Lamarckian inheritance.

Our positive emotions are not instinctual—at least not yet. They are not wired in us like our base instinctual emotions; they don't give rise to automatic responses. If only they did! I have come to realize that our next step of evolution consists of developing positive emotions as instincts. How do we do that?

Developing Positive Emotions

Remember emotions are physiological and psychological effects of feelings. They consist of the thoughts and judgments and the physiological effects, such as facial expressions, that arise in a particular feeling state. We cannot control our feelings or physiology so easily, so mystics through the ages have mostly talked about controlling the thoughts associated with the feelings. In the spiritual traditions, this effort is called developing virtue.

Suppose you are experiencing the feeling of anger in response to a certain bully. Your body tenses and angry thoughts arise: "Who does he think he is!" "I'll show him!" and so on. Mystics say, "Cool it!" Cool your anger by cooling your thoughts, replacing the indignation with love. Jesus said, "Love your enemy." Upon seeing your enemy, anger and violent feelings arise, along with associated thoughts of hate. If, instead of thinking hate, you think love, the negative feelings will also subside. The psychologist Uma

Krishnamurthy teaches that this approach is quite general: Whenever any negative emotion arises, we should treat it with a dose of positive emotion.

Alas! This approach has its problems. First, it is not easy to think loving thoughts when an enemy is confronting you. Instinctual behavior always has an element of helplessness. When threatened by an enemy, you cannot help but think hate. Second, even if you have a little control over your thoughts, before you act upon them you would like (especially if you are male) to have a guarantee that your loving thoughts are reciprocated by the enemy. What if the enemy interprets your nonviolence as weakness? Doesn't that make the situation worse?

Mystics remind us that we should give up such bargaining and instead practice "unconditional" love. If we keep on practicing this attitude in a relaxed way, eventually we may take a quantum leap to supramental intelligence. Then we will *know* for ourselves, with complete certainty, that loving thought is the right approach to handle negative feelings that arise in response to someone else's negative feeling.

Unfortunately, even with that quantum leap, the behavior change is guaranteed only in the context in which the quantum leap occurred. In a sufficiently different context, thoughts will show signs of negativity once again, and we will require more practice to transcend negative with positive thoughts, and more quantum leaps. In your entire life, you may take just a few such quantum leaps—if you are fortunate.

There is also a different kind of change, a transformative journey at the culmination of which the positive emotion will arise without effort. Uma Krishnamurthy says that after this transformation, instead of mind evaluating feelings, the evaluation is instantly referred to supramental intelligence, with quite different results. Unfortunately, the techniques that lead to such transformation are just too demanding to be applied universally (see chapter 17). The capacity to respond emotionally from the supramental level alone is extremely rare in human history. Maybe Buddha had it, and Jesus, but very few others.

The problem is that the supramental is not represented in us physically, so we cannot live completely from that level. Instead, we live in the mind that is represented in the brain. When negative feelings arise, the brain's memory associations spew off negative conditioned thoughts in the mind, giving rise to negative emotions. We have no permanent way of changing the pattern under all situations, short of developing the capacity of effortlessly quantum leaping to the supramental for all our emotional responses. Such development is a worthy goal but, unfortunately, one that isn't realistic for most people.

Some spiritual traditions, for example one called tantra, have suggested another approach: transforming the feelings directly, that is, by affecting the vital movements associated with feelings. This is the subject of raising kundalini, or bringing in new, creative vital energy—the positive transcending the negative—which I have discussed elsewhere (Goswami 2001). This technique is also much too rarefied to consider for achieving changes in humanity en masse.

Even if a few human beings do manage a permanent transformation through these techniques, is there any lasting gain for humanity? Certainly, those who transform in this way are the spiritual masters who originate our spiritual traditions, the great religions, for example, but the role of religions in transforming the world has been dubious at best. Are there other ways of taking our next evolutionary step?

EVOLVING POSITIVE EMOTIONS AS INSTINCTS

To understand the path forward in evolving emotions, we need to return to the crucial issue of the vital body and its morphogenetic fields. I mentioned the wiring in the limbic brain that makes a negative emotion into an instinct. What makes that wiring? Consciousness does, with crucial help from modifications in the vital morphogenetic fields associated with the limbic brain.

Every time a member of a species has a feeling in one of its first three chakras, the conditioning of its correlated vital morphogenetic field at that chakra is reinforced and the associated organs are affected. The organs are connected to the brain by two-way "psychoneuroimmunological" connections. So when we feel vital movement at a low chakra, through the psychoneuroimmunological connection from the associated organ to the limbic center, the feeling affects and conditions the vital morphogenetic fields not only of the chakra but also of the limbic brain. This conditioning is then mapped as a circuit of the limbic brain.

In animals (and also in primitive humans before language developed), individual consciousness was secondary to species consciousness. For the vital body, this relationship remains true to some extent even in humans, at least in our present state of evolution, explaining how we so readily acquire someone else's negative emotions. Through this species consciousness, vital conditioning is available to the entire species for use via reincarnation. With each generation, the probability is greater that the negative feeling conditioning of the vital morphogenetic field of the limbic brain will be used to wire the limbic brain of more and more members of the species. Eventually, through this process, the negative emotions become instincts (see chapter 20 for more on the creation of instincts).

That, I propose, is how we became beings awash in negative emotions. Where do we start to change the situation? Let's start with the individual. Every time I transform a negative emotion to a positive one by quantum leaping, the morphogenetic fields operating on the limbic center of my brain would change once again, probabilistically, of course. The more quantum leaps I take, the more the vital changes take root, and the morphogenetic field becomes conditioned to respond in the new way. In time, my limbic brain would develop new brain circuits corresponding to my positive feelings.

Note that I am talking about my personal vital body, not yours, and certainly not that of humanity. However, as I noted before, the vital individuality is weak, with no strong boundaries between where one person's vital body ends and where another's begins. In other words, there may still be some leftover role for the species consciousness here. Additionally, for emotions, it is natural to work on them not by oneself, but in relationship. So any transformative experience resonates at least in two people. Further, I can actively intend that my transformative experience be shared by all people, that the benefits of the quantum leaps accrue to all people.

Do you see how these ideas might apply to evolving the instinctual patterns of our species to include positive emotions? First, we consciously invoke species consciousness in our personal transformative journeys. Second, we participate in transformational work on emotions in bigger and bigger groups. This we do first in two-way, intimate relationships—lover and beloved, therapist and client, parent and child. Then we work in families. And finally we work in groups, such as in a business setting or in an educational context. Third, we use the power of intention to propagate the effect. In a matter of a few generations, limbic brain centers tuned for positive feelings can be inherited by the bulk of the population via reincarnational transfer of the modified morphogenetic fields. Genetic assimilation via biological creativity completes the job of making positive feelings instinctual.

This scenario is a major departure from how we traditionally think of spiritual work and practices. First of all, as soon as we utter the word *spirituality*, most people think of meditation—watching one's thoughts in solitude. Meditation is great for individual transformation, because the challenge is to transcend the strong mental ego and make room for God-consciousness. But even if we make brain programs of the mental wisdom gained from mental quantum leaps to the supramental, these programs will never get into the gene pool because species consciousness does not exist at the mental level. So by our usual spiritual work, the species will never change.

We have to face the fact that we have ignored the vital body for too long. The consequences are showing up in a major way in the form of mind-body disease (see Goswami 2004). Certainly, as a society we are opting for relative emotional isolation, because we just cannot handle emotions any more. It is much easier to have a purely mental relationship with another on the Internet. The only exceptions we allow are in relation to sex and in response to others' negative emotions, thanks to their strong instinctual character. All this has to change.

Transpersonal psychologists are right in emphasizing love over fear, positive over negative emotions. But the early thinking was too optimistic: that love is our natural being, that we will get to love just by somehow taking the fear away from our being (Jampolsky 2004). It is more complicated than that. But it is not impossible, not at all.

Can you imagine a world in which human beings naturally, instinctually relate with cooperation and love, not competition and envy? Now you can. There is some evidence that we have a couple of positive emotional instincts already in the form of universal limbic brain circuits. First, scientists report that if a certain center in the limbic brain is excited (by certain rituals), one gets an experience of God (see Newberg, D'Acquili, and Rause 2001 for a review of the data). The materialist tendency is to say this study shows that the God experience of humanity is entirely mechanical, due to a brain circuit developed in us through Darwinian evolution. (The rebuttal to this argument is a no-brainer, because instincts cannot be explained by Darwinism; see chapter 20). Because of this materialist denigration of God, I sometimes jokingly call this "God-in-the-brain" circuit the new G spot! But joking aside, the data must be reinterpreted as evidence that a certain kind of positive emotion of the sacred has become an instinct for all humanity, and this is why God won't go away. Our brains know about God.

Second, more recent research has uncovered a similar circuit in the limbic brain connected with certain types of altruism (Harbaugh, Mayr, and Barghart 2007). The presence of this circuit explains why altruism, at least in some simple form, is so universal, a human instinct.

If we have done it once, we can do it again. The evolutionary movement of consciousness is going to take us in that direction anyway, but we can choose to accelerate that movement.

GETTING HUMANITY UNSTUCK

To summarize, the next step of our evolution is on hold because the limbic brain has developed the instinctual circuits of negative emotions, and we have not learned how to deal with that. We have to change the vital fields correlated with the limbic brain once again to develop additional instincts of positive emotion, instincts that will guide the behavior of humanity en masse. Here is the key to getting humanity unstuck: We each go on striving for quantum leaps to the supramental response to emotion, but we use the occasion of the leaps not only in the journey of own transformation but also in the journey of evolution for humanity.

We have gone this route before—trying to develop positive emotions en masse, although never as instincts. Buddha tried it with great success during his lifetime. As soon as he passed away, however, the nonviolence movement ended, and violence came back in a hurry. Likewise, following Jesus' teachings, the early Christians tried to establish love as the reigning emotion en masse. They got nowhere as a religion until they co-opted the violent ways of the Roman Empire for spreading their teachings.

In spirituality, the means and the end cannot be contradictory, and Christianity has never recovered from this misalliance with empire.

Can we heed the lessons of a new evolutionary paradigm of science when we did not heed the lessons of Buddha and Jesus? We do gain one advantage by going it the scientific way: We have a better understanding of the underlying dynamics.

There is no need to institute a religion; the role of any religion, even a scientific one, is limited in this day and age. There is no need to co-opt any government to spread this science either. Instead, the evolutionary movement of consciousness will spread among active participants but outside the purview of the powers that be—unseen, almost unknown.

My vision for our evolutionary future is this: Let those who can, see the point of the new science. Let those who can, take quantum leaps from negative to positive emotions with evolutionary intentions. Let those who can, live increasingly with positive emotions, making new brain circuits and changing the associated morphogenetic fields. Let those who can, spread positive emotions through relationships. We will be few at first, but our numbers will grow, especially as we create new institutions that facilitate this journey for others.

Around us, the world will go on, for a while yet, in its violent ways: terrorism, violent reactions to terrorism, opportunistic curbing of democracy, greedy economics, proselytizing in the name of religion, destruction of our ecosystem. Meanwhile, the modified morphogenetic fields we have created through our personal journeys will be available for future generations to draw upon through reincarnation. More and more babies, though born to violent parents, will have positive emotional brain circuits. One day the making of positive emotional brain circuits will be canalized. Eventually, genetic assimilation will take place.

How long will this take? Not long. We are drawing on powerful processes: the evolutionary movement of consciousness, synchronicity, the living archetypes. Most importantly, we are drawing on the lives of our countless ancestors, who have already prepared the morphogenetic fields for the changes that only we, reincarnating them now, can make.

REFERENCES

Aspect, A., J. Dalibard, and G. Roger. 1982. Experimental test of Bell's inequalities with time varying analyzers. *Physical Review Letters* 49:1804–6.

Augros, R., and G. Stanciu. 1988. *The New Biology: Discovering the Wisdom in Nature*. Boston: Shambhala, New Science Library.

Aurobindo. 1996. *The Life Divine*. Pondicherry, India: Sri Aurobindo Ashram.

Barrow, J. D., and F. G. Tipler. 1986. *The Anthropic Cosmological Principle*. New York: Oxford Univ. Press.

Bass, L. 1971. The mind of Wigner's friend. *Harmathena*, no. 122. Dublin: Dublin Univ. Press.

Bateson, G. 1980. *Mind and Nature*. New York: Bantam.

Behe, M. J. 1996. *Darwin's Black Box*. New York: Simon & Schuster.

Bergson, H. 1949. *Introduction to Metaphysics*. New York: Harper.

Birch, C. 1999. *Biology and the Riddle of Life*. Sidney, Australia: Univ. of New South Wales Press.

Blood, C. 1993. On the relation of the mathematics of quantum mechanics to the perceived physical universe and free will. Rutgers University, preprint.

———. 2001. *Science, Sense, and Soul*. Los Angeles: Renaissance Books.

Boden, M. 1990. *The Creative Mind*. New York: Basic Books.

Bohm, D. 1980. *Wholeness and Implicate Order*. London: Rutledge & Kegan Paul.

Brough, J. 1958. Time and evolution. In *Studies in Fossil Vertebrates*, ed. T. S. Westoll. London: University of London/Athlone Press.

Brown, G. S. 1977. *Laws of Form*. New York: Dutton.

Bush, G. L. 1975. Modes of Animal Speciation. *Annual Review of Ecology* 6:339–64.

Cairns, J., J. Overbaugh, and J. H. Miller. 1988. The origin of mutants. *Nature* 335 (8 September):142–45.

Capra, F. 1982. *The Turning Point*. New York: Simon & Schuster.

———. 1996. *The Web of Life: A New Scientific Understanding of Living Systems*. New York: Doubleday.

Carroll, S. B. 2005. *Endless Forms Most Beautiful*. New York: Norton.

Casti, J. L. 1989. *Paradigms Lost: Images of Man in the Mirror of Science*. New York: Avon.

Chalmers, D. 1995. *Toward a Theory of Consciousness*. Cambridge, MA: MIT Press.

Chopra, D. 1990. *Quantum Healing*. New York: Bantam-Doubleday.

Darwin, C. 1859. *On the Origin of Species by Means of Natural Selection or the Preservation of Favored Races in the Struggle for Life*. London: Murray.

Davies, P. 1988. *The Cosmic Blueprint: New Discoveries in Nature's Creative Ability to Order the Universe*. New York: Simon & Schuster.

———. 1999. *The Fifth Miracle: The Search for the Origin and Meaning of Life*. New York: Simon & Schuster.

Dawkins, R. 1976. *The Selfish Gene*. New York: Oxford Univ. Press.

Dembski, W. A. 2002. *No Free Lunch: Why Specified Complexity Cannot Be Purchased without Intelligence*. New York: Rowman & Littlefield.

Dennett, D. C. 1995. *Darwin's Dangerous Idea*. New York: Simon & Schuster.

Devall, W., and G. Sessions. 1985. *Deep Ecology*. Salt Lake City, UT: G. M. Smith.

Dossey, L. 1992. *Meaning and Medicine*. New York: Bantam.

Durr, H. P., and F. T. Gottwald, eds. 1997. *Rupert Sheldrake in der Diskussion. Das Wagnis einer neuen Wissenschaft des Lebens (Gebundene Ausgabe)* [Scientists discuss Sheldrake's theory about morphogenetic fields]. Munich: Scherz.

Eden, D. 1999. *Energy Medicine*. New York: Tarcher/Putnam.

Efron, R. 1977. Biology without consciousness—and its consequences. In *Logic, Laws and Life: Some Philosophical Complications*, ed. R. G. Colodny, 209–233. Pittsburgh: Pittsburgh Univ. Press.

Eigen, M., and P. Schuster. 1979. *The Hypercycle: A Principle of Natural Self-Organization*. New York: Springer-Verlag.

Eigen, M., W. Gardiner, P. Schuster, and R. Winkler-Oswatitsch. 1981. The origin of genetic information. *Scientific American* 244:88–118.

Eldredge, N. 1985. *Time Frames*. New York: Simon & Schuster.

Eldredge, N., and Gould, S. J. 1972. Punctuated equilibria: An alternative to phyletic gradualism. In *Models of Paleontology*, ed. T. J. M. Schopf, 82–115. San Francisco: Freeman.

Elsasser, W. M. 1981. Principles of a new biological theory: A summary. *Journal of Theoretical Biology* 89:131–50.

———. 1982. The other side of molecular biology. *Journal of Theoretical Biology* 96:67–76.

Gish, D. T. 1978. *The Fossils Say No*. San Diego, CA: Creation-Life Publishers.

Goldschmidt, R. 1952. Evolution as viewed by one geneticist. *American Scientist* 40:84–123.

Goleman, D. 1995. *Emotional Intelligence*. New York: Bantam.

Goodwin, B. 1994. *How the Leopard Got Its Spots: The Evolution of Complexity*. New York: Charles Scribner's Sons.

Goswami, A. 1989. The idealist interpretation of quantum mechanics. *Physics Essays* 2:385–400.

———. 1991. *Quantum Mechanics*. Dubuque, IA: Wm. C. Brown.

———. 1993. *The Self-Aware Universe: How Consciousness Creates the Material World*. New York: Tarcher/Putnam.

———. 1994. *Science within Consciousness: Developing a Science Based on the Primacy of Consciousness*, IONS Research Report CP-7. Sausalito, CA: Institute of Noetic Sciences.

———. 1997a. Consciousness and biological order: Toward a quantum theory of life and evolution. *Integrative Physiological and Behavioral Science* 32:75–89.

———. 1997b. A quantum explanation of Sheldrake's morphic resonance. In *Rupert Sheldrake in der Diskussion. Das Wagnis einer neuen Wissenschaft des Lebens (Gebundene Ausgabe)* [Scientists discuss Sheldrake's theory about morphogenetic fields], ed. H. P. Durr and F. T. Gottwald. Munich: Scherz.

———. 1999. *Quantum Creativity*. Cresskill, NJ: Hampton Press.

———. 2000. *The Visionary Window: A Quantum Physicist's Guide to Enlightenment*. Wheaton, IL: Quest Books.

———. 2001. *Physics of the Soul*. Charlottesville, VA: Hampton Roads.

———. 2002. *The Physicists' View of Nature, Part 2: The Quantum Revolution*. New York: Kluwer/Plenum.

———. 2004. *The Quantum Doctor*. Charlottesville, VA: Hampton Roads.

———. 2008. *God Is Not Dead*. Charlottesville, VA: Hampton Roads.

Goswami, A., and L. M. Simpkinson. 1999. *A new science of dreams*. Unpublished manuscript, Institute of Noetic Sciences, Sausalito, CA.

Goswami, A., and D. Todd. 1997. Is there conscious choice in directed mutation, phenocopies and related phenomena? An answer based on quantum measurement theory. *Integrative Physiological and Behavioral Science* 32:132–42.

Gould, S. 1980. Is a new and general theory of evolution emerging? *Paleobiology* 6:119–30.

Gould, S. J., and N. Eldredge. 1977. Punctuated equilibria; the tempo and mode of evolution reconsidered. *Paleobiology* 3:115–151.

Gould, S. J., and R. C. Lewontin. 1979. The spandrels of San Marco and the Panglossian paradigm. *Proceedings of the Royal Society of London*, B 205:581–98.

Grad, B. 1964. A telekinetic effect on plant growth. *International Journal of Parapsychology* 6:472–98.

———. 1965. Some biological effects of "laying-on of hands": A review of experiments with animals and plants. *Journal of the American Society for Psychical Research* 59:95–127.

Grad, B., R. G. Cadoret, and G. I. Paul. 1961. The influence of unorthodox methods of treatment on wound healing in mice. *International Journal of Parapsychology* 3:5–24.

Grant, V. 1977. *Organismic Evolution*. San Francisco: Freeman.

Gregory, R. L. 1981. *Mind in Science*. London: Weidenfeld and Nicholson.

Grinberg-Zylberbaum, J., M. Delaflor, L. Attie, and A. Goswami. 1994. Einstein Podolsky Rosen paradox in the human brain: The transferred potential. *Physics Essays* 7:422–8.

Gruber, H. 1981. *Darwin on Man*. 2nd ed. Chicago: Univ. of Chicago Press.

Hamilton, W. D. 1964. The genetical theory of social behavior. Pts. 1 and 2. *Journal of Theoretical Biology* 7:1–16; 17–32.

Harbaugh, W. T., U. Mayr, and D. R. Barghart. 2007. Neural responses to taxation and voluntary giving reveal motives for charitable donations. *Science* 316 (15 June):1622–5.

Harman, W., and H. Reingold. 1984. *Higher Creativity*. Los Angeles: Tarcher.

Harman, W., and E. Sahtouris. 1998. *Biology Revisioned*. Berkeley, CA: North Atlantic Books.

Harrison, J. 1996. How Ludwig Became a Homunculus. *Philosophy* 71(277):439–44.

Hellmuth, T., A. G. Zajonc, and H. Walther. 1986. Realization of "delayed-choice" experiments. In *New Techniques and Ideas in Quantum Measurement*

Theory, ed. D. M. Greenberger, 108–114. New York: New York Academy of Science.

Hitching, F. 1982. *The Neck of the Giraffe*. New York: New American Library.

Ho, M. W. 1993. *The Rainbow and the Worm*. Singapore: World Scientific.

Ho, M. W., and P. Saunders, eds. 1984. *Beyond Neo-Darwinism: An Introduction to the New Evolutionary Paradigm*. London: Academic Press.

Hobson, J. A. 1990. Dreams and the brain. In *Dreamtime and Dreamwork*, ed. S. Krippner. New York: Tarcher / Perigee.

Hofstadter, D. R. 1980. *Gödel, Escher, Bach: An Eternal Golden Braid*. New York: Vintage Books.

Huxley, J. 1960. The evolutionary vision. In *Issues in Evolution*, vol. 3 of *Evolution after Darwin* (The University of Chicago Centennial Discussions), ed. S. Tax and C. Callendar, 249–61. Chicago, IL: Univ. of Chicago Press.

Jablonka, E., and M. J. Lamb. 2005. *Evolution in Four Dimensions*. Cambridge, MA: MIT Press.

Jampolsky, G. G. 2004. *Love is Letting Go of Fear*. Berkeley, CA: Celestial Arts.

Jung, C. G. 1971. *The Portable Jung*, ed. J. Campbell. New York: Viking.

Kaufman, S. 2002. *Unified Reality Theory: The Evolution of Existence into Experience*. Milwaukee, WI: Destiny Toad Press.

Kimura, M. 1983. *The Neutral Theory of Molecular Evolution*. Cambridge, MA: Cambridge Univ. Press.

Kirschner, M. W., and J. C. Gerhart. 2005. *The Plausibility of Life*. New Haven, CT: Yale Univ. Press.

Koestler, A. 1978. *The Case of the Midwife Toad*. N.p.: Hutchinson.

Krishnamurthy, U. 2008. *Yoga Psychology*. In press.

Lad, V. 1984. *Ayurveda: The Science of Self-Healing*. Santa Fe, NM: Lotus Press.

Laszlo, E. 2004. *Science and the Akashic Field*. Rochester, VT: Inner Traditions.

Lewis, C. S. 2002. *Poems*. New York: Harcourt Trade Publishers, Harvest Books.

Lewontin, R. 2000. *The Triple Helix*. Cambridge, MA: Harvard Univ. Press.

Libet, B., E. Wright, B. Feinstein, and D. Pearl. 1979. Subjective referral of the timing of a cognitive sensory experience. *Brain* 102:193.

Lipton, B. 2005. *The Biology of Belief*. Santa Rosa, CA: Mountain of Love / Elite Books.

Lovelock, J. 1982. *Gaia: A New Look at Life on Earth.* Oxford: Oxford Univ. Press.

MacGregor, G. 1978. *Reincarnation in Christianity.* Wheaton, IL: Theosophical Publishing House, Quest Books.

Mahadevon, T. M. P. 2000. *Upanishads.* New Delhi: Motilal Banarsidass.

Marcel, A. 1980. Conscious and preconscious recognition of polysemous words: Locating the selective effect of prior verbal contexts. In *Attention and Performance VIII,* ed. R. S. Nickerson, 435–57. Hillsdale, NJ: Lawrence Erlbaum.

Margulis, L. 1993. *Symbiosis in Cell Evolution: Microbial Communities in the Archean and Proterozoic Eons.* New York: Freeman.

Margulis, L., and D. Sagan. 1986. *Microcosmos: Four Billion Years of Evolution from Our Microbial Ancestors.* New York: Simon & Schuster.

Maslow, A. H. 1971. *The Farther Reaches of Human Nature.* New York: Viking.

Maturana, H. 1970. Biology of cognition. Reprinted in H. Maturana and F. Varela, eds. 1980. *Autopoiesis and Cognition.* Dordrecht, Holland: D. Reidel.

Maturana, H., and F. Varela. 1980. *Autopoiesis and Cognition.* Dordrecht, Holland: D. Reidel.

Mayr, E. 1942. *Systematics and the Origin of Species.* New York: Columbia Univ. Press.

———. 1963. *Animal Species and Evolution.* Cambridge, MA: Belknap Press of Harvard Univ. Press.

———. 1980. The evolution of Darwin's thought. *Guardian* (London), July 22, 18.

———. 1982. *The Growth of Biological Thought.* Cambridge, MA: Harvard Univ. Press.

McCarthy, K., and A. Goswami. 1993. CPU or self-reference? Can we discern between cognitive science and quantum functionalist models of mentation? *Journal of Mind and Behavior* 14:13–26.

Miller, K. R. 1999. *Finding Darwin's God.* New York: HarperCollins.

Miller, S. 1974. The first laboratory synthesis of organic compounds under primitive conditions. In *The Heritage of Copernicus: Theories "Pleasing to the Mind,"* ed. J. Neyman, 228–42. Cambridge, MA: MIT Press.

Minkoff, E. G., and P. G. Baker. 2004. *Biology Today: An Issues Approach.* New York: Garland Publishers.

Mitchell, M., and A. Goswami. 1992. Quantum mechanics for observer systems. *Physics Essays* 5:525–9.

Moura, G., and N. Don. 1996. Spirit possession, Ayahuasca users and UFO experiencers: Three different patterns of states of consciousness in Brazil. Abstract of talk presented at the 15th International Transpersonal Association Conference, May 16–21, 1996, Manaus, Brazil. Mill Valley, CA: International Transpersonal Association.

Newberg, A., E. D'Acquili, and V. Rause. 2001. *Why God Won't Go Away.* New York: Ballantine.

Page, C. 1992. *Frontiers of Health.* Saffron Walden, UK: C. W. Daniel.

Pelletier, K. 1992. *Mind as Healer, Mind as Slayer.* New York: Delta.

Penfield, W. 1976. *The Mystery of the Mind.* Princeton, NJ: Princeton Univ. Press.

Penrose, R. 1989. *The Emperor's New Mind.* Oxford: Oxford Univ. Press.

Pert, C. 1997. *Molecules of Emotion.* New York: Scribner.

Prigogine, I. 1980. *From Being to Becoming.* San Francisco: Freeman.

Prigogine, I., and I. Stengers. 1984. *Order Out of Chaos.* New York: Bantam.

Ridley, M. 1996. *The Origins of Virtue.* New York: Penguin.

Ring, K., and S. Cooper. 1995. Can the blind ever see? A study of apparent vision during near-death and out-of-body experiences. University of Connecticut, preprint.

Robinson, H. J. 1984. A theorist's philosophy of science. *Physics Today* 37:24–32.

Russell, E. S. 1930. *The Interpretation of Development and Heredity.* Oxford: Clarendon Press.

Sabel, A., C. Clarke, and P. Fenwick. 2001. Intersubject EEG correlations at a distance—The transferred potential. In *The 44th Convention of the Parapsychological Association: Proceedings of Presented Papers,* ed. C. S. Alvarado, 419–22. Raleigh, NC: Parapsychological Association.

Sabom, M. 1981. *Recollections of Death: A Medical Investigation.* New York: Harper & Row.

Sancier, K. M. 1991. Medical applications of Qigong and emitted qi on humans, animals, cell cultures, and plants. *American Journal of Acupuncture* 19:367–77.

Saunders, P. 1980. *An Introduction to Catastrophe Theory.* New York: Cambridge Univ. Press.

Schmidt, H. 1993. Observation of a psychokinetic effect under highly controlled conditions. *Journal of Parapsychology* 57:351–72.

Schrödinger, E. 1944. *What is Life?* Cambridge: Cambridge Univ. Press.

Schroeder, G. L. 1997. *The Science of God.* New York: Broadway Books.

Searle, J. R. 1987. Minds and brains without programs. In *Mindwaves: Thoughts on Intelligence, Identity and Consciousness*, ed. C. Blakemore and S. Greenfield. Oxford: Basil Blackwell.

———. 1994. *The Rediscovery of the Mind.* Cambridge, MA: MIT Press.

Shapiro, R. 1987. *Origins: A Skeptic's Guide to the Creation of Life on Earth.* New York: Summit Books.

Sheldrake, R. 1981. *A New Science of Life.* Los Angeles: Tarcher.

———. 1999. *Dogs That Know When Their Owners are Coming Home.* New York: Three Rivers Press.

Simpson, G. 1944. *The Tempo and Mode of Evolution.* New York: Columbia Univ. Press.

Smythies, J. R. 1994. *The Walls of Plato's Cave: The Science and Philosophy of Brain, Consciousness and Perception.* Aldershot, UK: Avebury.

Spiegelman, S. 1967. An *in vitro* analysis of a replicating molecule. *American Scientist* 55:3–68.

Standish, L. J., L. Kozak, L. C. Johnson, and T. Richards. 2004. Electro-encephalographic evidence of correlated event-related signals between the brains of spatially and sensory isolated human subjects. *The Journal of Alternative and Complementary Medicine* 10:307–14.

Stapp, H. P. 1993. *Mind, Matter, and Quantum Mechanics.* New York: Springer.

Steele, E. 1980. *Somatic Selection and Adaptive Evolution.* New York: State Mutual Book.

Stevenson, I. 1974. *Twenty Cases Suggestive of Reincarnation.* Charlottesville: Univ. Press of Virginia.

———. 1977. Research into the evidence of man's survival after death. *Journal of Nervous and Mental Disease* 165:153–83.

———. 1987. *Children Who Remember Previous Lives: A Question of Reincarnation.* Charlottesville: Univ. Press of Virginia.

Strohman, R. 1992. *The forces of life.* Univ. of California, Berkeley, preprint.

Taylor, G. R. 1983. *The Great Evolution Mystery.* New York: Harper & Row.

Teilhard de Chardin, P. 1961. *The Phenomenon of Man.* New York: Harper & Row.

Thewissen, J. G. M., S. T. Hussain, and M. Arif. 1994. Fossil evidence for the origin of aquatic locomotion in archaeocete whales. *Science* 263:210–12.

Thom, R. 1975. *Structural Stability and Morphogenesis.* Reading, MA: Benjamin.

Ullman, D. 1988. *Homeopathy: Medicine for the 21st Century*. Berkeley, CA: North Atlantic Books.

Vithulkas, G. 1980. *The Science of Homeopathy*. New York: Grove Press.

Von Neumann, J. 1955. *The Conceptual Foundations of Quantum Mechanics*. Princeton: Princeton Univ. Press.

———. 1966. *The Theory of Self-Reproducing Automata*. Urbana, IL: Univ. of Illinois Press.

Wackermann, J., C. Seiter, H. Keibel, and H. Walach. 2003. Correlation between brain electrical activities of two spatially separated human subjects. *Neuroscience Letters* 336:60–64.

Waddington, C. 1957. *The Strategy of the Genes*. London: Allen & Unwin.

Wallace, R., and H. Benson. 1972. The physiology of meditation. *Scientific American*, 226:84–90.

Weil, A. 1995. *Spontaneous Healing*. New York: Knopf.

Weisman, A. 1893. *Die Allmacht der Naturzuchtung*. Jena: Gustav Fischer.

Weiss, R. 2005. Evolution debate in Kansas spurs battle over school materials, *Washington Post*, October 28.

Wesson, R. 1991. *Beyond Natural Selection*. Cambridge, MA: MIT Press.

Wheeler, J. 1975. The Universe as Home for Man. In *The Nature of Scientific Discovery: A Symposium Commemorating the 500th Anniversary of the Birth of Nicolaus Copernicus*, ed. O. Gingerich, 262–95. Washington, DC: Smithsonian Press.

———. 1977. Genesis and observership. In *Foundational Problems in the Special Sciences*, ed. R. E. Butts and J. Hintikka, 3–34. Dordrecht: D. Riedel.

Wilber, K. 1981. *Up from Eden*. Garden City, NY: Anchor/Doubleday.

———. 1996. *A Brief History of Everything*. Boston: Shambhala.

Winik, L. W. 2005. McCain in Defense of Darwin. *Parade*, October 30. http://www.parade.com/articles/editions/2005/edition_10-30-2005/intelligence_report_0 (accessed December 14, 2007).

Wolpert, L. 1993. *The Triumph of the Embryo*. New York: Oxford Univ. Press.

Wyller, A. 1999. *The Creating Consciousness*. Denver, CO: Divina.

———. 2003. Beyond Darwin's paradigm: Consciousness driving evolution. Unpublished manuscript.

Yanchi, Liu. 1988. *The Essential Book of Traditional Chinese Medicine*. New York: Columbia Univ. Press.

*I*NDEX

acquired characteristics
  alternatives to Darwinism,
    28–29, 192
  causal flows and, 142–43, 298
  inheritance of, 28, 305
  Lamarckism, 297–98, 300
actualities, 22
acupuncture, 48, 231–34
adaptation, 18, 202, 298
adenine, 89
Age of Enlightenment, 275
Age of Mammals, 207
"Aha!" experience, 68
Allen, Woody, 46
allopathic medicine, 232–34
alternative medicine. *See* homeopathy
altruism, 315
ambulacral systems of echinoderms,
  163
Ambulocetus natans, 190
amino acids, 89–90, 93
anthropic principle, 110–13
anthropomorphism, 229–30
archaeopteryx, 184
arrow of time, 4, 24–25, 69
Arthur (computer program), 99
Aspect, Alain, 59
Atlas of Evolution (de Beer), 183
atomic nuclei, 111–12
attractors, 99, 187–88, 215, 231
Augros, Robert, 174, 195, 229
auras, 237
Aurobindo, Sri, 6, 72, 79–80
autocatalysis, 86, 92
autopoiesis, 99, 103

autoscopic vision, 285–86
awareness, 73, 246. *See also* consciousness
ayurveda, 48, 213, 235

Backster, Cleve, 238–39
bacteria, 29
Baker, P. G., 86
Balinese culture, 188
Baptiste de Monet, Jean, 28
Barrow, J. D., 110–11
Bass, Ludwig, 40
Bateson, Gregory, 103
Behe, Michael, 12, 110, 113, 126,
  153, 216–17
Belousov-Zhabotinsky Reaction,
  91–92
Bergson, Henri, 48, 213
Berkeley, Bishop, 261–62
biblical beliefs, 11–12, 21, 261–62
big bang theory, 107–8
biochemical form building, 55, 61,
  211–12, 216
biological creativity, 69
biological forms. *See also* molecular
  biology; organismic biology
  biochemical machine example,
    55, 61, 211–12
  biology, incompleteness of, 47–50
  consciousness and, 49, 174
  creativity and, 51
  depth psychology, 49
  deterministic laws, 51
  development of, 8–9, 216–18,
    269–70

328

Dembski, William, 101–2
Dennett, D., 147, 158
Descartes, René, 42, 59, 74, 141, 287
deterministic physics, 139–40
Devall, W., 320
dialectic schools, 154
Dirac, Paul, 23
directed mutation, 29, 192
direct perception model, 261–64
discontinuity, 22, 34, 36
DNA (deoxyribonucleic acid molecules), 9, 88f–89, 145
DNA/RNA protein conglomerate, 94
Dobzhansky, Theodore, 303
Don, Norman, 288
doshas (defects), 235–36
Dossey, Larry, 320
double-helix structure. *See* DNA
dream analysis, 253–54
Driesch, Hans, 170–71
Drosophilia (fruit flies) experiment, 218
dualism. *See also* world views
  collapse events and, 37
  defined, 13
  dynamics and, 184–88
  problem of, 22
  resolution of, 59, 66
Durr, H. P., 320

ecology. *See* ethics; selfish gene
Eden, D., 320
Efron, Robert, 54
ego, 33, 309
ego-consciousness, 34
Eigen, Manfred, 92, 95, 103
Einstein, Albert, 107
élan vital, 48, 216
Eldredge, N., 7–8, 19, 21, 150
Elsasser, W. M., 23, 181, 235
embryo, 17, 211, 214
embryology, 17, 298

emotional intelligence, 308
emotional maturity, 308
emotions. *See also* chakras
  as brain epiphenomena, 227–28
  as instincts, 311–15
  cancer and, 255
  contagiousness of, 308
  defined, 222
  mind and body and, 222–23
  molecules of emotion, 222
  negative, 308–12
  of animals and plants, 238–40
  positive, 309–11
  sexual drives, 309
  transformation of, 311
Emperor's New Mind, The (Penrose), 252
engrams, 253
entities, 54
entropy law of, 24, 42, 180–81
enzymes, 89–91
epigenetic inheritance systems, 300
epileptic patients, and memory, 253
epiphenomenalism, 144
epistemological techniques, 54
ethics. *See also* selfish gene
  bioethics, 82, 146, 291–93
  deep ecology, 293–95
  evolution and, 292
  religious, 291–92
  scientific, 292
eukaryotic cells, 87, 162, 173
evo-devo, 154, 217–18
evoked potential signal, 41
Evolutionary Hymn (Lewis), 78
evolutionary issues, 10, 19, 44
evolutionary stages, 77–78, 270–71
evolution, future of, 307–8, 316–17
evolutionism, 15, 19, 25, 36, 51
evolution, purpose of, 72
evolvability, 147, 153–54
external forces, and physical change, 139